Springer Series in Optical Sciences

Volume 221

Springer Series in Optical Sciences is led by Editor-in-Chief William T. Rhodes, Georgia Institute of Technology, USA, and provides an expanding selection of research monographs in all major areas of optics:

– lasers and quantum optics
– ultrafast phenomena
– optical spectroscopy techniques
– optoelectronics
– information optics
– applied laser technology
– industrial applications and
– other topics of contemporary interest.

With this broad coverage of topics the series is useful to research scientists and engineers who need up-to-date reference books.

More information about this series at http://www.springer.com/series/624

Stephan Stuerwald

Digital Holographic Methods

Low Coherent Microscopy and Optical
Trapping in Nano-Optics and Biomedical
Metrology

 Springer

Stephan Stuerwald
University of California, Berkeley
Berkeley, CA, USA

ISSN 0342-4111   ISSN 1556-1534   (electronic)
Springer Series in Optical Sciences
ISBN 978-3-030-00168-1   ISBN 978-3-030-00169-8   (eBook)
https://doi.org/10.1007/978-3-030-00169-8

Library of Congress Control Number: 2018954606

This Springer imprint is published by the registered company Springer Nature Switzerland AG
The registered company address is: Gewerbestrasse 11, 6330 Cham, Switzerland

Preface

A variety of applications in the area of biomedicine, microchemistry, and micro system technology demand the possibility of a minimally invasive, positioning or manipulation and analysis of biological cells or different micro and nano particles, which is not possible with conventional mechanical methods. To achieve these requirements, systems are established that facilitate a micromanipulation with light. Therefore, a momentum change of photons that are refracted on the surface of a transparent cell or a partly transparent micro particle is harnessed for holographic optical tweezers (HOTs), which enables the user to exert forces in the range of piconewtons.

An additional challenge is the exact and detailed imaging of biological cells in a bright field microscope. The outline of a cell is usually easily recognizable due to diffraction or absorption characteristics, but the reconstruction of a three-dimensional form is not directly possible. A suitable method to overcome this problem is the imaging with digital holographic quantitative phase contrast microscopy (DHM). This method utilizes the optical path difference of photons that pass through a transparent cell with a higher refractive index compared to the ambient medium and can therefore be considered as an advanced and quantitative phase contrast microscopy method.

A combination of these two methods is so far solely rudimentary explored in research. A further developed expansion of these functionalities into an integrated setup therefore represents a significant enhancement in this field and allows new applications. A few examples for new applications are an automated detection of any biological cell and a subsequent automated holding, separation, analysis of rheological cell parameters, and a positioning into specifically predetermined allocated cavities for cell sorting, as needed in Lab-on-a-Chip (LoC) applications (Fig. 1b).

The aim of this work is the development and characterization of a system for analyzing biological as well as technical specimens with digital holographic methods. The system shall allow a realization and characterization of the combination of both described techniques with unrivaled high content cell analysis and manipulation capabilities. To achieve this, a confocal laser scanning and fluorescence

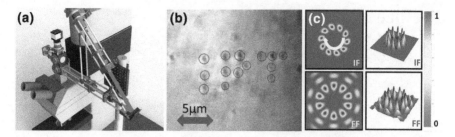

Fig. 1 **a** Developed multifunctional microscope system in CAD illustration with beam paths indicated in red. **b** 14 micro particles arranged by dynamic holographic optical tweezers to the letters "IPT". **c** Intermediate (IF) and far-field (FF) intensity distributions of Bessel beams of higher order (5th) for generating symmetric trap patterns and optical angular momentum

microscope is extended: The camera port of a microscope beam path is utilized for coupling in a modulated laser beam besides imaging a hologram of the object plane onto a camera. To enable, e.g., a simultaneous manipulation of multiple cells ($\gg \approx 6$) in three dimensions and in video rate, an intensity distribution in the object plane is created by an optical reconstruction of a hologram. To achieve this, a spatial light modulator (SLM) that can change the phase of the wavefront in every pixel in the range of $[0, 2\pi]$ is inserted into the beam path. This phase pattern gets Fourier transformed optically by a microscope objective, leading to the desired intensity distribution in the focus plane of the microscope objective.

This method is based on a phase-only SLM and facilitates also the generation of non-diffractive beam configurations like Airy, Mathieu, and Bessel beams of higher order, which allow for efficient, stable, and mostly symmetric trap configurations. It is demonstrated that they open up prospect for increased light efficiency for direct laser writing (DLW). Non-diffractive beams are sometimes referred to *self-healing* beams, since self-focusing effects lead to better focusing properties exceeding the conventional Rayleigh length compared to standard Gaussian beams. Additionally, it is shown that these beams and their higher modes enable an allocation of a high number of different efficient light force and light intensity distributions. These can be particularly useful for a three-dimensional positioning of micro and nano particles or structures and even for direct laser writing.

The development of modular systems for DHM and HOTs also allows an integration into a nano positioning system (NMM-1, $25 \times 25 \times 5$ mm^3, positioning accuracy: ≈ 3 nm) in order to extend the positioning volume beyond the field of view and depth of sharpness of the microscope objective and more precise than with conventional microscopy xyz-stages.

The digital holographic imaging mode is additionally investigated with low coherent light sources like SLDs, LEDs, and a super continuous light source to optimize the phase noise of the interferograms. The system combination is then applied for direct laser writing in photoresists based on two-photon polymerization —also known as 3D-lithography. This opens up the possibility for dynamic and multifocal generation of large-scale micro and nano structures without drawbacks in accuracy caused by stitching losses.

Previous systems solely allow direct laser writing with one fixed focus point and a scan volume of typically $300 \times 300 \times 300$ m^3 with a xyz-piezo stage. The combined deployment of bigger nano positioning systems like the NMM-1 ($25 \times 25 \times 5$ mm^3) or the recently new developed NPMM-200 ($200 \times 200 \times 25$ mm^3) open up the possibility to realize a photonic system platform for direct production, manipulation, assembly and measuring of photonic circuits and elements.

Therefore, this work contributes to the long-term goal of establishing a modular photonic system platform with multifunctional features, which represent a key technology for the efficient production of micro and nano optical structures. These system platforms shall also open the way to new research areas in the field of the systems itself and provide a future tool for the whole nano optic and photonic area.

Berkeley, USA   Stephan Stuerwald

Contents

Chapter 1
Introduction

Abstract Real-time high-throughput identification, high content screening, characterisation and processing of reflective micro and nano structures as well as (semi-)transparent phase objects like biological specimens are of significant interest to a variety of areas ranging from cell biology and medicine to lithography. In this thesis, the two optical techniques, namely digital holographic microscopy (DHM) and holographic optical tweezers (HOTs) are realised in one integrated setup that permits to minimal-invasively image, manipulate, control, sense, track and identify micro and nano scaled specimens or even fabricate them with direct laser writing in three dimensions. The setup allows to operate both methods simultaneously additionally to conventional bright field, confocal laser scanning (LSM) and fluorescence microscopy. This enables e.g. time lapse experiments in microfluidic environments with stably arranged or distanced cells respectively. Furthermore the combination of both holographic methods allows to examine and monitor e.g. cell deformations, their volume or reactions to certain artificially manipulated surface types as well as apoptosis and the influence radius of released messengers. For a better signal to noise ratio, different low coherent light sources including supercontinuum light sources are investigated. The system setup and its functionality are characterised in this work. Holographic optical tweezers not only open up the possibility for generating multiple dynamic traps for micro and nano particles, but also the opportunity to exert optical torque with special complex electromagnetic fields like Bessel beams, which can facilitate the movement and rotation of particles. Further non diffracting beams are investigated for utilisation in the holographic optical tweezer (HOT) system for self-assembling of micro particles which comprises Mathieu, Laguerre and Airy beams. New opportunities that arise from the complex methods are investigated also for direct laser writing of nano structures based on two photon polymerisation.

Introduction to Investigated Holographic Methods

Today microscopy, electron or optical, with wavelengths ranging from infrared to EUV (Extreme Ultra Violet) and X-rays, plays an essential role in the development of micro- and nanotechnologies such as photolithography, MEMS and MOEMS (Micro Optical, Electronical and Mechanical Systems). Their use in material and surface sciences is also in strong progression. Nowadays the perspective to reach

© Springer Nature Switzerland AG 2018
S. Stuerwald, *Digital Holographic Methods*, Springer Series
in Optical Sciences 221, https://doi.org/10.1007/978-3-030-00169-8_1

the nanometre scale with optical super resolution systems, that are easier to handle than electron microscopes, represents a growing incitation to develop new technological approaches. Digital Holographic Microscopy (designated with DHM in the following) represents one of them. In life science, there is an increased interest in low-cost minimal-invasive and high-throughput methods for manipulation, identification, as well as characterisation of biological micro/nano organisms like cells and tissue. Optical methods are seen as a key technology for areas like next generation point-of-care health solutions, cancer diagnosis, food safety, environmental monitoring and early detection of pandemics. Despite their morphologically simple and minute nature, biological micro and nano organisms exhibit complex systems showing a sophisticated interaction with their environment. Usually, practices in biological investigations have been dominated by bio-chemical processes which are typically preparation intensive, time consuming and invasive or even toxic. Among the range of technologies aiming in this direction, optical methods often permit a balanced alternative of minimal-invasiveness, low cost, sensitivity, speed and compactness. In conventional microscopy, semi-transparent cells need to be fixed, stained or labelled with fluorescent markers for characterizing them thoroughly, which may even kill the cells or adversely affect their viability caused by their partly toxic nature, disrupting the cells' natural life cycle. Thus further investigations are significantly influenced in certain measurement applications such as stem cell screening and time lapse experiments. In order to annihilate several of these restrictions, three-dimensional (3D) microscopy techniques have been investigated [1, 2]. In particular, technological advances have permitted increased use of coherent optical systems [3] mainly due to availability of a wide range of inexpensive laser sources, advanced detector arrays as well as spatial light modulators. For further information on holography and it's development throughout the time, please refer to Appendix B.2.1.

Up to the present a variety of holographic systems for microscopic applications has been developed for optical testing and quality control of reflective and (partially) transparent samples [4–7] (see also Fig. 1.1). Combined with microscopy, digital holography permits a fast, non-destructive, full field, high resolution and minimal invasive quantitative phase contrast microscopy with an axial resolution better than 5 nm, which is particularly suitable for high resolution topography analysis of micro and nano structured surfaces as well as for marker-free imaging of biological specimen [6, 7].

Digital holography is a branch of the imaging science, which deals with numerical reconstruction of digitally recorded holograms and with computer synthesis of holograms and diffractive optical elements. Basically, this means that in addition to the intensity the phase of the electromagnetic field is recorded or manipulated. In digital holography, the hologram, formed by a superposition of coherent object and reference waves, is recorded with a sensor (e.g., a CCD camera), that converts the intensity distribution of the incident light into an electrical signal. After a subsequent discretisation, the information is stored with a computer for digital post processing. The reconstruction of a signal wave is performed with the aid of numerical reconstruction algorithms. In case of holographic optical tweezers (denoted with "HOT"), a pre-calculated phase pattern is generated with a spatial light modulator (type: liquid

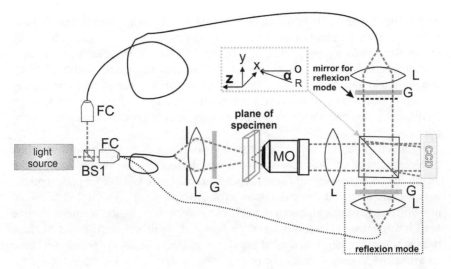

Fig. 1.1 Schematic setup of different interferometric configurations for application on transparent and reflective specimen. L: lenses; G: attenuation filter; MO: microscope objective, FC: fiber coupler; BS: beam splitter

crystal on silicon, LCOS) which alters a wavefront to such a shape that in the focus plane of a microscope objective a desired intensity pattern is created.

Digital Holographic Microscopy is in principle an imaging method offering both real time observation capabilities and sub-wavelength resolution [5, 8]. The reconstruction of a complex wavefront from a hologram is giving the amplitude and the absolute phase of a light wave altered by microscopic objects. In general, an absolute phase contrast offers axial accuracies usually better than 10 nm, one nanometre in air or even less in dielectric media [8]. The lateral accuracy and the corresponding resolution can be kept diffraction limited by utilisation of a high numerical aperture (NA) microscope objectives (MO). In the present state of the art, it can be kept commonly below 600 nm when operating in the visible range and without combination of new super-resolution microscopy techniques like 4π-microscopy, TIRF, thresholds, localisation microscopy (STORM, (F)PALM), model fit methods (image restoration, model based imaging) and nonlinear imaging (STED, RESOLFT, super resolving masks, resist nonlinearity - e.g. double patterning).

Content of Book in Brief

In this work, the principles of hologram formation, acquisition and wavefront reconstruction from digitally captured holograms, acquired in a non-scanned modality, are first described and then further developed in detail with regard to innovative approaches for microscopic imaging and trapping of micro and nano scaled objects. This comprises the following aspects, which are briefly summarised:

In the framework of this work, a setup is realised which allows a simultaneous and independent operation of the digital holographic imaging mode and the holographic trapping technique. Thus time lapse experiments in microfluidic environments can be performed with stably arranged or distanced cells, which allows to examine e.g. their deformations, changes in volume or reactions to certain drugs or artificially manipulated surface types as well as apoptosis and the influence radius of released messengers. Holographic optical tweezers not only open up the possibility of generating multiple dynamic traps for micro and nano particles, but also the opportunity to exert optical torque with special beam configurations like Bessel beams that can facilitate the movement and rotation of particles or even generate optically induced pumps.

Structured light fields have therefore increased significance in optical trapping, manipulation and organisation. Non-diffracting beams feature a magnified Rayleigh length compared to standard Gaussian beams and thus are suitable for optical potentials that are extended along the beam axis. Till now a variety of different optical non-diffracting beams and optical fields carrying orbital angular momentum have been investigated. This covers Laguerre beams, Bessel beams, Airy and Mathieu beams, which can carry an optical angular momentum of $n\hbar$ per photon and have an azimuthal angular dependence of $\exp(in\phi)$, where n denotes the index for the unbounded azimuthal mode and ϕ is the azimuthal angle. Optical beams which offer an angular momentum give prospect to versatile applications such as arranging and rotating particles for biomedical or chemical applications. Here, it is also put emphasis on higher-order non-diffractive Bessel beams and the generation of superimposed higher-order Bessel beams.

Higher order Bessel beams may be superimposed in such a way that they produce a field which either has or does not have a global optical momentum. When generating a superimposed higher-order Bessel beam, a rotation in the field's intensity profile as it propagates is clearly demonstrated, but which may result in no global angular momentum in specific axial planes. The superposition of light fields is encoded on a spatial light modulator (SLM) with a ring-slit hologram (which may optionally also be generated by illumination of an axicon with an overlay of Laguerre-Gaussian beams). Here, both the near- and far-field intensity profiles of a ring-slit aperture generated with a liquid crystal on silicon (LCOS) spatial light modulator are investigated. For harnessing these special beams for holographic optical traps, knowledge of the phase and amplitude distribution of the complex beam at various planes in the area of the sharp image plane is required.

Because of their possible functionality for the arrangement of micro particles and structures, further types of non-diffracting beams are discussed as well as their application for optical manipulation. For example the utilisation of Mathieu beams and combined parabolic and Airy beams as stable beam types for arranging particles and their distances in a desired geometry is demonstrated. The experimental realisation of these extraordinary laser beams is as well performed with a spatial light modulator (LCOS-type). These also named *non-diffracting beams* are relatively immune to diffraction and additionally, they exhibit transverse acceleration while propagating. These extraordinary properties of the demonstrated beams can facilitate the arrangement of both micro particles and biologic cells within a region of interest of a microfluidic sample chamber through induced particle transport along curved trajectories.

In contrast, the Airy beam spot is built up of a bright main spot and several side lobes whose intensity increases towards the main spot. Therefore, micro particles and cells experience a gradient force which drags them into the main spot. Due to the light pressure exerted, micro particles and cells are then levitated and propelled along the curved trajectory of the main spot away from the cleared region.

Due to the possibility of complex wavefront retrieval, digital holographic methods allow also altering the focus numerically by propagating the complex wave. Especially for compensation of deformations or displacements and for long-term investigations of living cells, a reliable region selective numerical readjustment of the focus is of particular interest in digital holographic microscopy and has therefore been treated in this work. Since this method is time consuming, a Halton point set with low discrepancy has been chosen. By this refocusing, the effective axial resolution and depth of field shall be enhanced numerically by post processing of complex wave fronts without narrowing the field of view leading to a loss of information around the focus plane by blurring. The concept of numerical parametric lenses is another key feature in DHM and used to correct aberrations in the reconstructed wave front caused by the setup. To reduce the number of parameters for these parametric lenses in comparison to a standard Zernike polynomial basis ad in order to make parameter adjustments more intuitive, the polynomial basis by Forbes is applied for the needs of DHM. Both numerical approaches are characterised and adapted to the requirements of DHM.

Partial coherent light sources open up prospect for phase noise reduction in digital holographically reconstructed phase distributions by suppressing multiple reflections in the experimental setup [7, 9]. Thus, superluminescent diodes (SLDs) and supercontinuum light sources are investigated for application in digital holographic microscopy. Besides the spectral properties and the resulting coherence lengths of the utilised light sources are characterised, an analysis of dispersion effects and their influences on the hologram formation is carried out experimentally and theoretically with adaption of approaches from optical coherence tomography (OCT). The low coherence limits the maximum interference fringe umber in off axis holography for spatial phase shifting. Thus, the application of temporal phase shifting based digital holographic reconstruction techniques is considered as well. It is demonstrated that the low coherent light sources lead to a reduction of noise in comparison to a laser light based experimental setup.

The applicability of the mentioned methods is demonstrated by results of investigations of micro and nano structured surfaces as well as biological cells. Applications to cell dynamics studies like nano-movements and cyto-architectures deformations are also shown.

References

1. Liebling, M., Blu, T., Cuche, E., Marquet, P., Depeursinge, C., Unser, M.: A novel non-diffractive reconstruction method for digital holographic microscopy. In: Proceedings of the 2002 IEEE International Symposium on Biomedical Imaging, pp. 625–628 (2002)

2. Liebling, M., Blu, T., Unser, M.: Complex-wave retrieval from a single off-axis hologram. J. Opt. Soc. Am. A **21**(3), 367–377 (2004). https://doi.org/10.1364/JOSAA.21.000367
3. Gleeson, M.R., Sheridan, J.T.: A review of the modelling of free-radical photopolymerization in the formation of holographic gratings. J. Opt. A Pure Appl. Opt. **11**(2), 024008 (2009). http://stacks.iop.org/1464-4258/11/i=2/a=024008
4. Marquet, P., Rappaz, B., Magistretti, P.J., Cuche, E., Emery, Y., Colomb, T., Depeursinge, C.: Digital holographic microscopy: a noninvasive contrast imaging technique allowing quantitative visualization of living cells with subwavelength axial accuracy. Opt. Lett. **30**(5), 468–470 (2005). https://doi.org/10.1364/OL.30.000468
5. Mann, C., Yu, L., Lo, C.-M., Kim, M.: High-resolution quantitative phase-contrast microscopy by digital holography. Opt. Express **13**(22), 8693–8698 (2005). https://doi.org/10.1364/OPEX.13.008693
6. Charrière, F., Kühn, J., Colomb, T., Montfort, F., Cuche, E., Emery, Y., Weible, K., Marquet, P., Depeursinge, C.: Characterization of microlenses by digital holographic microscopy. Appl. Opt. **45**(5), 829–835 (2006). https://doi.org/10.1364/AO.45.000829
7. Martínez-León, L., Pedrini, G., Osten, W.: Applications of short-coherence digital holography in microscopy. Appl. Opt. **44**(19), 3977–3984 (2005). https://doi.org/10.1364/AO.44.003977
8. Marquet, P., Rappaz, B., Charrièrec, F., Emery, Y., Depeursinge, C., Magistretti, P.: Analysis of cellular structure and dynamics with digital holographic microscopy. In: Biophotonics 2007: Optics in Life Science, vol. 6633. Optical Society of America (2007)
9. Dubois, F., Joannes, L., Legros, J.-C.: Improved three-dimensional imaging with a digital holography microscope with a source of partial spatial coherence. Appl. Opt. **38**(34), 7085–7094 (1999). https://doi.org/10.1364/AO.38.007085

Chapter 2
Theory

This chapter comprises an introduction into the most significant theoretical backgrounds of digital holography. For a better understanding, also the basics of conventional, classic holography including temporal and spatial phase shifting techniques are summarized at the beginning of this chapter, before proceeding with the numerical propagation of complex object waves and special considerations that are required for application in a microscope system. Further, different types of spatial light modulators for a complex manipulation of electromagnetic waves are introduced and discussed. Several approaches for their utilization in a microscope system are then introduced. These include aberration control, focusing possibilities and the exertion of a momentum for single or multiple holographic optical traps (HOTs). Furthermore, dynamic holography for optical micromanipulation in life science microscopy and different applications of optical tweezers are theoretically discussed. As a significant topic in latest research, so-called diffractive and non-diffractive beam types are introduced comprising Bessel, Mathieu and Airy beams. In a last part of this chapter, the basics of direct laser writing with two-photon polymerization are explained which can be improved by utilization of spatial light modulators.

2.1 Basic Principles of Holography

Since the invention of holography in 1947 by the physicist Dénes Gábor [1] constantly new applications are being developed that exploit the principle of holography. Particularly after the development of the laser in 1960 significant progress has been achieved since for the first time a coherent and monochromatic light source was available whose coherence properties are required for holography. Holography is an imaging method in which not only the intensity (amplitude) and the wavelength of a light image is recorded - such as it is the case for a photograph or colour photography - but in addition to the amplitude the phase distribution is detected in the image field. Since the phase cannot be measured directly, it is detected indirectly via the camera

© Springer Nature Switzerland AG 2018
S. Stuerwald, *Digital Holographic Methods*, Springer Series
in Optical Sciences 221, https://doi.org/10.1007/978-3-030-00169-8_2

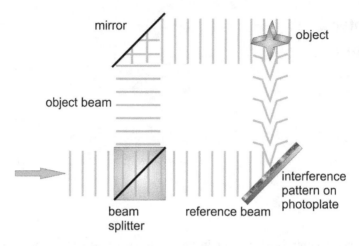

Fig. 2.1 Schematic setup for hologram recording: a coherent wave is split up into a reference and an object beam. Both waves interfere and form a hologram on the recording medium

based recording and evaluation of interferograms. Above all, the wave nature of light is utilised [2].

In general, holography is therefore a method for recording and reconstruction of complex wavefronts. In order to perform holographic recording of an object, it is illuminated with coherent light in transmission or reflection geometry. By coherent superposition of a reference wave $\vec{E}_R(\vec{r}, t)$ with an object wave $\vec{E}_O(\vec{o}, t)$ that has been diffracted, reflected or scattered at an object, the interferogram $I_H(\vec{o}, \vec{r})$ in the hologram plane is formed and recorded with analogue or digital media (Fig. 2.1). This interferogram can be expressed by (2.4) [3–5], where k denotes the (scalar) wave number, ω the angular frequency, t the time and $\vec{o}$, $\vec{r}$ the respective position vector.

$$\vec{E}_O(\vec{o}, t) = \vec{E}_{0O}(\vec{o})\ e^{i(\omega t - k_o o - \phi_o)}, \tag{2.1}$$

$$\vec{E}_R(\vec{r}, t) = \vec{E}_{0R}(\vec{r})\ e^{i(\omega t - k_r r - \phi_R)}, \tag{2.2}$$

$$I_H(\vec{o}, \vec{r}) = |\vec{E}_O(\vec{o}, t) + \vec{E}_R(\vec{r}, t)|^2 \tag{2.3}$$

$$= |\vec{E}_O(\vec{o})|^2 + |\vec{E}_R(\vec{r}, t)|^2 + \vec{E}_O(\vec{o}, t) \cdot \vec{E}_R^{\,*}(\vec{r}, t) + \vec{E}_R(\vec{r}, t) \cdot \vec{E}_O^{\,*}(\vec{o}, t)$$

$$= |\vec{E}_O(\vec{o})|^2 + |\vec{E}_R(\vec{r}, t)|^2 + 2|\vec{E}_O(\vec{o})||\vec{E}_R(\vec{r})| \cos(\phi_O(\vec{r}) - \phi_R(\vec{r}))$$

$$= I_O(\vec{r}) + I_R(\vec{r}) + 2\sqrt{I_O(\vec{r}) I_R(\vec{r})} \cos(\phi_O(\vec{r}) - \phi_R(\vec{r})). \tag{2.4}$$

The parameters I_O, I_R and ϕ_O, ϕ_R denote the intensity and phase of the object and reference wave respectively. In contrast to conventional photography, additionally to the amplitude the phase of the wave front is saved in a medium allowing a spatial impression of an object. In a second step the complex wave originated from the object can be reconstructed entirely with the help of a reference wave or numerical methods in case of digital holography. In the following section, a differentiation between classic and digital (numerical, discretised) holography is specified.

2.1.1 Classic Holography

In case of a classic, optical reconstruction the hologram is illuminated with the same type of reference wave utilised for the recording process, resulting in a modulation of the reference wave E_R in the hologram plane (x, y) with the transmittance $\tilde{T}$. Neglecting the dependencies in time and the vectorial character, the following equation for the transmitted field strength applies [3–5]: $E_T(x, y) = E_R(x, y) \cdot \tilde{T}(x, y)$.

In case of an amplitude hologram, the amplitude transmission $\tilde{T}$ is a variable of the impinged energy $B(x, y) = \int_0^{t_E} I(x, y, t)dt$ during the exposure time t_E and can be approximated in a linear domain of the recording medium [3–5] (e.g. a photographic film):

$$\tilde{T}(x, y) = a - bt_E I_H(x, y) \qquad (2.5)$$

Here the parameters a and b are real constants describing a linear function. The formed spatial amplitude modulation constitutes the hologram. The spatial resolution is determined by the wavelength, the angle between the object and the reference wave as well as the specific properties of the recording medium [6]. In general, it has to be taken into consideration that the spatial resolution of the hologram restricts the size of the object to be recorded.

For the optical reconstruction, in general the arrangement of the reference wave and the hologram is retained and the hologram is illuminated solely with the reference wave (Fig. 2.2). By this, in the direct vicinity of the hologram an electromagnetic field E_T is generated which is equal to the product of the amplitude transmittance $\tilde{T}$ of the medium and impinged field intensity E_R [3–5] that is proportional to the energy:

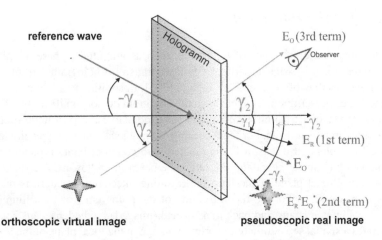

Fig. 2.2 Reconstruction of an "off-axis"-hologram with spatial separation of the images

$$E_T(x, y) = E_R(x, y) \cdot \tilde{T}(x, y) = aE_R(x, y) - bt_E I_H(x, y)$$
$$= aE_R(x, y) - bt_E\big(E_R(x, y)|E_R(x, y)|^2 + E_R(x, y)|E_O(x, y)|^2 +$$
$$+ E_R(x, y)E_O^*(x, y)E_R(x, y) + E_R(x, y)E_O(x, y)E_R^*(x, y)\big) \qquad (2.6)$$
$$= aE_R(x, y) - bt_E\big(\underbrace{E_R(x, y)\,(I_R(x, y) + I_O(x, y))}_{\text{Zero order}}$$

$$+ \underbrace{E_O(x, y)I_R(x, y)}_{\text{virtual image}} + \underbrace{E_R^2(x, y)E_O^*(x, y)}_{\text{real image}}\big) \qquad (2.7)$$

The first term in (2.7) is proportional to the irradiated reference wave and thus contains an offset with no relevant information for the reconstruction of the image (*zero order*). The second term represents the *virtual, orthoscopic* (non inverted) image, whose electromagnetic field is proportional to the original object wave. Therefore, the object seems to be located at the same place where it has been positioned during the recording process (Fig. 2.2). The third part of the equation is proportional to the conjugated complex wave front and gives a real, pseudoscopic image where the depths are inverted. This mathematical constituent is denoted as *twin image*. In case of parallel vectorial directions of object and reference wave in the hologram plane, the arrangement is named as *in-line* holography. Then, all the three terms are superposed and cannot be separated easily. In contrast, this can be avoided by application of *off-axis* holography [6] where the object and reference wave are slightly tilted towards each other with an angle α. In this case, all terms in (2.7) are separated spatially during reconstruction (Fig. 2.2, [7, 8]). Furthermore, this overlap of the different terms is also avoided if the so-called non-diffractive reconstruction methods are applied (see also Sect. 2.1.3).

2.1.2 Fourier Holography

Of the many different types of holograms such as amplitude, phase, or rainbow holograms, for this work in particular Fourier transform holograms are of special importance as they offer several advantages [3–5]. In addition to a relatively simple mathematical description, local errors in the hologram do not significantly affect the quality of the reconstruction since for each image point in the reconstruction plane the entire hologram is utilised. For the recording of an object, which is typically located in the focal plane of a lens, it is illuminated with coherent light. Typically, a lens collimates this object beam and the reference beam in an "off-axis" arrangement. In the other focal plane of the lens, the resulting interference pattern is recorded on a film (Fig. 2.3). For the reconstruction of the hologram it is then illuminated with a collimated beam, and back transformed (imaged) by the lens. This approach generates a spatial separation of the primary image in the focal plane of the lens, the zero diffraction order and the conjugate image.

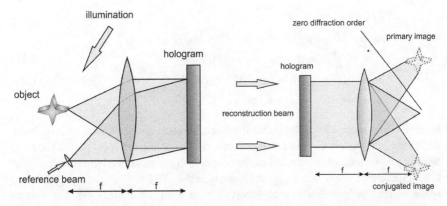

Fig. 2.3 Recording (left) and reconstruction (right) of a Fourier transform hologram in off-axis geometry [3]

If the complex amplitude emanating from the object plane is described with $E_O(x, y)$, then the complex amplitude of the wave from the hologram plate in the focal plane of the lens corresponds to its Fourier transform (FT) [3]:

$$\epsilon_O(\xi, \eta) = \mathscr{F}\{E_O(x, y)\}.$$ (2.8)

The reference wave is a point source whose origin is located in the front focal plane at the position (x_0, y_0). The complex amplitude of the reference wave in the hologram plane is given by

$$\epsilon_R(\xi, \eta) = e^{-i2\pi(\xi x_0 + \eta y_0)}.$$ (2.9)

The transmission hologram formed in the hologram plane by the interference of the two waves is described by the following equation:

$$\begin{aligned} T(\xi, \eta) &= a - bt_E(\epsilon_O + \epsilon_R)(\epsilon_O + \epsilon_R)^* \\ &= a - bt_E(\epsilon_R\epsilon_R^* + \epsilon_O\epsilon_O^* + \epsilon_R^*\epsilon_O + \epsilon_R\epsilon_O^*)(\xi, \eta). \end{aligned}$$ (2.10)

Where the parameters a, b denote real material constants describing a linear function, t_E the exposure time (illumination duration) and ϵ_O the Fourier transformed object wave. For the reconstruction of the hologram it is illuminated with the collimated reference wave from (2.9), which provides the following relation:

$$T\epsilon_R = (a - bt_E|\epsilon_R|^2)\epsilon_R - bt_E(\epsilon_O\epsilon_O^*\epsilon_R + \epsilon_R\epsilon_R^*\epsilon_O + \epsilon_R\epsilon_R\epsilon_O^*).$$ (2.11)

If this collimated beam containing the hologram information is finally imaged by a lens, the FT is formed in the focal plane of the lens $T\epsilon_R$ [6].

$$\mathscr{F}\left\{T\epsilon_R\right\}(x', y')$$
$$= (a - bt_E |\epsilon_R|^2)\mathscr{F}\left\{\epsilon_R\right\}(x', y') - bt_E(\mathscr{F}\left\{\epsilon_O \epsilon_O^* \epsilon_R\right\}$$
$$+ \mathscr{F}\left\{\epsilon_R \epsilon_R^* \epsilon_O\right\} + \mathscr{F}\left\{\epsilon_R \epsilon_R \epsilon_O^*\right\})(x', y') \qquad (2.12)$$
$$= C_1 \delta(x' + x_0, y' + y_0) - C_2[(E_O \otimes E_O)(x' + x_0, y' + y_0)$$
$$+ E_O(-x', -y') + E_O^*(x' + 2x_0, y' + 2y_0)]$$

Here, the first term with a Delta distribution represents the zero order of diffraction and the second term typically corresponds to a ring around the focal point. The third term is proportional, but inverted to the original object wave, and the fourth term is the conjugate of the original object wave shifted by $(-2x_0, -2y_0)$. Both last terms result in real images that can be recorded with a digital camera [3]. For a more detailed introduction see Appendix B.2.

For a better illustration, in the following the direct relationship between the hologram and reconstruction plane is considered. Therefore the illuminated hologram from (2.11) is designated with $H(\xi, \eta)$ and the reconstruction of this hologram in the object plane with $h(x', y')$. As mentioned above, the mathematical relationship between these planes is given by the FT. The optical lens performs this FT between front and back focal plane according to:

$$h(x', y') = \mathscr{F}\left\{H(\xi, \eta)\right\}$$
$$= \int \int_{-\infty}^{\infty} H(\xi, \eta) \exp\left[-i\frac{2\pi}{\lambda f}(x'\xi + y'\eta)\right] d\xi d\eta. \qquad (2.13)$$

Here, the amplitude and phase at the position (x', y') in the reconstruction plane is given by the amplitude and phase of the two spatial frequencies $(f_\xi = x'/\lambda f, f_\eta = y'/\lambda f)$ in the hologram plane [9].

The majority of the most used holograms is represented by amplitude holograms, especially due to their ease of preparation (e.g., exposure of a film). The modulation is then carried out in a reconstruction by different absorption in the transilluminated hologram medium. This represents a disadvantage of amplitude holograms as the absorption leads to an optical power loss of the reconstructed beam. By modulation of the phase, no loss of performance is induced. These holograms are often realised by harnessing a difference in the propagation time (optical path difference (OPD)) of different material thickness or refractive index. Another major advantage of phase holograms is a better quality in the reconstruction of the hologram. Thus it is theoretically possible by a so-called "blazed phase grating" - also known as *sawtooth* - to direct approximately the entire incident light radiation in the first diffraction order [10].

2.1.3 Digital Holography

The principles of hologram formation, acquisition and wave front reconstruction from digitally captured holograms, acquired in a non-scanned modality, are described here in more detail. In contrast to classic holography, where a phase hologram on a photosensitive material is recorded, in digital holography an amplitude hologram is captured with a digital camera and is thus directly available as a matrix of grey values. The digital reconstruction of the object wave is then performed numerically [11–13]. For digital holographic recording, usually CCD (Charged Coupled Device) or CMOS (Complementary Metal Oxide Semiconductor)-sensors are utilised, whereat the latter often exceed the performance of CCDs regarding sensitivity and and noise since several years. The exposure time is adjusted in such a way that the dynamic range of the semiconductor device is exploited in an optimal way.

Thus the hologram recording is comparable to an analogue recorded pure amplitude hologram with linear characteristic of the holographic film. With digital image acquisition, the characteristic response is not inverted like for instance with silver halide photo plates. Therefore, one can calculate - in first order - with an approximated real part of the amplitude transmittance $\tilde{T}$, which means $\tilde{T}(x_n, y_m) \propto I_H(x_n, y_m)$.

Thereby, the parameters (x_n, y_m) denote the discrete surface coordinates of the image capturing sensor surface with a pixel resolution of $N_x \times N_y$ and with t_E for the exposure time. A virtual "illumination" of the transmission $\tilde{T}(x_n, y_m)$, resultant from the intensity distribution in the camera plane containing the reference wave E_R yields [6]:

$$E_T(x_n, y_m) = E_R(x_n, y_m) \cdot \tilde{T}(x_n, y_m)$$
$$\approx E_R(x_n, y_m) \cdot I_H(x_n, y_m) \tag{2.14}$$
$$= \underbrace{E_R(x, y)(I_R(x, y) + I_O(x, y))}_{\text{zero order}} + \underbrace{E_O(x, y)I_R(x, y)}_{\text{virtual image}} \tag{2.15}$$
$$+ \underbrace{E_R^2(x, y)E_O^*(x, y)}_{\text{real image}}$$

Here, analogue to classic holography (see (2.7)) a partitioning of the transmitted wave E_T in three terms is resulting: Zero order, virtual and real image (2.14).

2.1.4 Computer Generated Holograms

The computer-generated holography (CGH) is a method that allows to calculate arbitrary holographic interference patterns on a computer. These generated holograms can be printed on a film, written with lithographic techniques, or displayed on a spatial light modulator (SLM). Due to the digital generation of the hologram, additional degrees of freedom are possible. This opens up the possibility to generate abstract and highly complex light distributions in the reconstruction plane that are otherwise not feasible in reality. When utilizing a SLM, dynamic light fields can be realised through changes in the hologram at typically 60 Hz to 1 kHz. The most significant advantage however is the encoding of only the desired terms from (2.7) to (2.12) in the hologram. Therefore disturbances like the zero order diffraction or the twin-image problem theoretically can be neglected [3].

The algorithms for hologram generation are numerous and are constantly being improved. Three main categories in which the algorithms can be classified are summarised in the following:

Iterative Fourier transformation algorithms (IFTA): The algorithms of this group are used when the intensity pattern consists of complex functions that cannot be described analytically. Since for many applications, only the intensity distribution in the reconstruction plane is of concern, the phase distribution represents an additional degree of freedom. The input variables are the desired intensity distribution in the reconstruction plane and a randomly selected phase distribution in the hologram plane. The computer propagates this randomly selected phase pattern to the reconstruction plane and replaces the resulting intensity distribution with the desired intensity distribution. After back-propagation to the hologram plane the amplitude and phase distribution are updated in order to approach to an optimum. With each iteration cycle the intensity distribution in the reconstruction plane gets closer to the desired intensity distribution [14] and usually converges to an optimum.

Direct search algorithms: In principle, this group of algorithms belongs also to the class of iterative algorithms. The desired intensity distribution in the reconstruction plane and a randomly selected phase distribution in the hologram plane serve as input variables. The latter is propagated to the reconstruction plane and then compared with the desired intensity pattern. The degree of equality is determined with a merit function.[1] In a next iteration the output hologram is varied randomly and after propagation another comparison is performed. If the new intensity pattern is closer to the desired value (the value of the corresponding merit function is smaller), the new hologram will retain and serve as the basis for the change in the next iteration. If the pattern is worse than the previous one, it will be discarded. Therefore an approximation to the desired intensity pattern over the course of the iteration process is made [15].

Analytical description of the hologram: If the desired intensity patterns in the reconstruction plane are relatively simple, analytically describable patterns (e.g., an array of $\{x, y, z\}$ focal points), the calculation of the corresponding hologram can be performed analytically. The advantage is a significantly faster calculation and less required computing power. Therefore, this method is especially useful when dynamic hologram generation is required. A detailed description of an exact algorithm can be found in the Appendix A.1.2 exemplarily for optical tweezers.

---

[1]The Merit function is a function that determines the correspondence between data and target model for a particular choice of parameters. The smaller the merit function the better the congruence.

2.1.5 *Numerical Reconstruction of Digital Holograms*

$$E_{\Delta z}(x', y') = \frac{1}{i\lambda} \int_{-\infty}^{\infty} \int_{-\infty}^{\infty} E_0(x, y) \frac{e^{\frac{i2\pi R}{\lambda}}}{R} \cos\theta \, dxdy$$

$$\text{with: } R = \sqrt{(x' - x)^2 + (y' - y)^2 + \Delta z^2} \tag{2.16}$$

Fundamental for a numerical reconstruction is the scalar diffraction theory of Kirchhoff with the derived equations from the Maxwell Kirchhoff's integral theorem (2.16) [9]. The corresponding theory allows the calculation of the complex amplitude $E(x', y')$ of a wave at any point B in the x', y'-plane (Fig. 2.4), if on any closed surface A around a point B the complex amplitude is known by its derivative.

At vertical incidence of the plane reference wave E_R onto the hologram in the xy-plane, the complex amplitude $E_{\Delta z}(x', y')$ at a distance Δz can be reconstructed by the hologram plane with the Fresnel–Kirchhoff's diffraction integral [6, 11] (2.16). The derived Huygens–Fresnel principle, according to the first Rayleigh–Sommerfeld solution in Cartesian coordinates, is shown in (2.17). For a detailed description of the scalar diffraction theory and the derivation by (2.17) it is referred to [3, 9] and the Appendix B.3.3, B.3.4.

$$E_{\Delta z}(x', y') = \frac{\Delta z}{i\lambda} \int_{-\infty}^{\infty} \int_{-\infty}^{\infty} E_0(x, y) \frac{e^{\frac{i2\pi R}{\lambda}}}{R^2} \, dxdy \tag{2.17}$$

$$\text{with: } R = \sqrt{(x' - x)^2 + (y' - y)^2 + \Delta z^2} \tag{2.18}$$

The complex amplitude $E_{\Delta z}(x', y')$ of a product B in the x', y'-plane results from the superposition of all - from the xy-plane outgoing - elementary waves with complex amplitudes $E_0(x', y')$ and the wavelength λ.

Equation (2.17) involves the inherent approximations of the scalar diffraction theory. Here, the vectorial character of light is neglected. The conditions for the validity

Fig. 2.4 Hologram planes x, y and x', y'

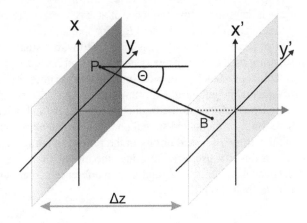

of this approximation are that, firstly, the diffracting structures must be larger than the wavelength. On the other hand, the diffracted field can not be considered too close to the aperture ($\Delta z \gg \lambda$) if a cosine approximation is utilised [9].

For the numerical evaluation of the integral (2.17) the so called Fresnel approximation is helpful to reduce the calculation effort. It comprises a replacement in (2.18) according to the given distance R in Cartesian coordinates in the argument of the exponential function and R^2 in the denominator of (2.17) by the first two terms respectively the first term of the Taylor series: $\sqrt{1+b} = 1 + \frac{1}{2}b - \frac{1}{8}b^2 + \cdots$ (see (2.19)). This corresponds to a paraboloidal approximation of the plane of the hologram starting from spherical waves. Thus from (2.17) results the following relation (2.20):

$$R \approx \Delta z \left(1 + \frac{1}{2} \left(\frac{x'-x}{\Delta z} \right)^2 + \frac{1}{2} \left(\frac{y'-y}{\Delta z} \right)^2 \right) \tag{2.19}$$

$$E_{\Delta z}(x', y') = \frac{e^{i\frac{2\pi \Delta z}{\lambda}}}{i\lambda \Delta z} \int_{-\infty}^{\infty} \int_{-\infty}^{\infty} E_0(x, y) \exp\left(i\frac{\pi}{\lambda \Delta z} \left((x'-x)^2 + (y'-y)^2 \right) \right) dxdy \tag{2.20}$$

2.2   Phase Shifting Reconstruction Methods

After an introduction into the general theory and some classical methods, a numerical realisation of the object wave reconstruction is described in from Sect. 2.1.3 on. Alternatively, the electric field of the object wave can be determined with the use of phase shifting methods. These diffraction-free methods avoid the occurrence of some limitations like the superposition of the zero diffraction order in an "in-line" configuration and the lower resolution in an "off-axis" arrangement. For this reason, only phase shifting methods are applied in this work.

In general, it has to be distinguished between spatial and temporal phase shifting methods [6, 16]. In temporal phase shifting, the phase and amplitude of an object wave is determined by a sequence of at least three interferograms of different relative phase relation between E_O and E_R. For spatially phase shifting methods the calculation of phase and amplitude is performed with a single interferogram, containing a superimposed carrier fringe system which represents a phase gradient. A subsequent calculation for each pixel takes into account the intensity of its adjacent pixels instead of temporal neighbouring interferograms. Another method consists in calculating the complex amplitude in the frequency domain [6, 17], which is not further discussed here. Both temporal and spatial phase shifting methods are utilised in this work and are therefore discussed briefly in the following sections.

For this it is useful to introduce the modulation $\gamma_0(x, y)$. It comprises a relation between the intensities I_R and I_O of a reference and an object wave respectively and is given by (2.21).

$$\gamma_0(x, y) \equiv \frac{2\sqrt{I_R(x, y)I_O(x, y)}}{I_R(x, y) + I_O(x, y)} \tag{2.21}$$

With $I_R(x, y) + I_O(x, y) = I_0(x, y)$, the insertion of (2.21) in (2.4) leads to the following interferogram equation in (2.22).

$$I_H(x, y) = I_0(x, y) \cdot \left(1 + \gamma_0(x, y) \cos\left(\varphi(x, y)\right)\right) \tag{2.22}$$

$$\text{with:} \quad \varphi(x, y) = \phi_0(x, y) + x\alpha_x + y\alpha_y + C \tag{2.23}$$

The spatial phase gradient induced by a tilt of a plane reference wave with respect to the object wave is expressed by $\vec{\alpha} = (\alpha_x, \alpha_y)$. The parameter C describes a constant global phase offset.

When digitizing a hologram of a continuous intensity distribution $I_H(x, y)$, it is discretised. Since individual detector elements (pixels) have a finite extent, a discretisation is carried out with an integration over the photosensitive area of each pixel of a recording sensor. In case of a pixel with coordinates (x_n, y_m) and an approximately constant spatial phase shift $\Delta\alpha$ between E_O and E_R, the effective modulation applied to a sensor with $N_x \times N_y$ pixels is given by [6, 16]:

$$\gamma(x_n, y_m) = \text{sinc}\left(\frac{\Delta\alpha}{2}\right) \frac{2\sqrt{I_R(x_n, y_m)I_O(x_n, y_m)}}{I_R(x_n, y_m) + I_O(x_n, y_m)} \tag{2.24}$$

$$= \text{sinc}\left(\frac{\Delta\alpha}{2}\right) \gamma_0.$$

The mean effective modulation $\bar{\gamma}$ is identical with the contrast. The systematic error during phase shifting using digital imaging sensors caused by the quantisation process are treated in [18, 19].

2.2.1 Temporal Phase Shifting

To calculate the relative phase and amplitude of every single pixel in temporal phase shifting interferometry, interferograms are recorded in which typically the reference wave has a different relative phase offset in each interferogram (e.g. constant shift of typically [30, 45, 60, 90, 120] degrees [16, 20]. For the reconstruction of the phase and the amplitude of the object wave, the discretely shifted interferograms respectively the intensity grey scale values of the image matrix are utilised for solving the interferogram equation (2.22). This results in an equation system with three unknown variables (I_0, γ, φ), which has a unique solution for at least three temporally shifted interferograms. Consequently, there exists a variety of algorithms, which are often named according to the number of required phase shifted intensity values I_j of recorded interferogram points $j \in (x_n, y_m)$. In this work, variable phase shift methods are also tested as they allow a variation of the phase shift α_t between the different interferograms and are therefore optimised for the specific experimental setup. A commonly used method is the variable three-step algorithm, whose derivation can be found in [6, 21]. By this algorithm shown in (2.25), the phase $\phi_{0,j}$ of the object wave of the jth interferogram can be calculated with a (to some extend always present, but not required) carrier fringe system generated by a phase gradient $x_n\alpha_x + y_m\alpha_y + C$. Furthermore, a variable five-step algorithm (2.26) is shown here.

$$(\phi_{0,j} + x_n\alpha_x + y_m\alpha_y + j\alpha_t + C) \bmod 2\pi = \arctan\left(\frac{(1 - \cos\alpha_t)I_{j-1} - I_{j+1}}{\sin\alpha_t(2I_j - I_{j-1} - I_{j+1})}\right) \tag{2.25}$$

$$(\phi_{0,j} + x_n\alpha_x + y_m\alpha_y + j\alpha_t + C) \bmod 2\pi = \arctan\left(\frac{(1 - \cos 2\alpha_t)I_{j-1} - I_{j+1}}{\sin\alpha_t(2I_j - I_{j+1} - I_{j-2})}\right) \tag{2.26}$$

The algorithms return the object phase, due to the ambiguity of the tangent function, solely modulo π. A demodulation by π to the interval 2π can be performed by taking into account the sign of the numerator and the denominator in the calculation of the inverse tangent function.

For calibration of the temporal phase step α_t of a corresponding mirror with attached piezo actuators as well as for an automatic test of the phase step before each measurement, (2.27) and (2.28) are utilised [16]:

$$\alpha_t = 2\arctan\sqrt{\frac{3(I_{j-1} - I_j) - (I_{j-2} - I_{j+1})}{(I_{j-1} - I_j) + (I_{j-2} - I_{j+1})}} \tag{2.27}$$

$$\alpha_t = \arccos\left(\frac{I_{j+2} - I_{j-2}}{2(I_{j+1} - I_{j-1})}\right) \tag{2.28}$$

In case the mirror - or respectively the plane - is explicitly shifted during a recording with a camera, a time integration occurs during the interferogram recording additional to the spatial integration over a pixel. This can be taken into account by a second sinc-term (see (2.22)):

$$I_H(x_n, y_m)_j = I_0(x_n, y_m) \cdot \Big(1 + \underbrace{\gamma_0(x_n, y_m)\mathrm{sinc}\frac{\Delta\alpha}{2}\mathrm{sinc}\frac{\Delta\alpha_t}{2}}_{=:\gamma(x_n, y_m)} \tag{2.29}$$

$$\cdot \cos\big(\phi_{0j}(x_n, y_m) + x_n\alpha_x + y_m\alpha_y + j\alpha_t + C\big)\Big)$$

The temporal, constant phase gradient between interferograms is denoted by α_t and the phase offset during recording by $\Delta\alpha_t$. The spatial phase shift, which is given by $x_n\alpha_x + y_m\alpha_y$, is not required here, but is also shown for completeness. For temporal phase shifting methods an "in-line" geometry is advantageous since the removal of the spatial phase background pattern - mainly consisting of a tilt and offset - can be omitted. By an explicitly discrete shift of the phase difference ("phase-stepping") the additional sinc-factor is theoretically omitted, too.

The accuracy of the temporal phase shifting methods is dependent on the intensity noise and the sensitivity towards the error of the phase step $\Delta\alpha_t$. This is described in [16, 22] and is minimised in the case of a three step algorithm for $\alpha_t = 90°$. The susceptibility to intensity noise of the different algorithms is investigated in [23, 24]. At the same time an object phase error, which is caused by the uncorrelated intensity noise, in dependence on the cross correlation of the intensity errors $< \Delta I_1 \Delta I_2 > \equiv \Delta I$, becomes minimal for 120°.[2] The error function $\sigma_{IR}(\Delta I, \gamma, I_0)$ for the three step algorithm is given in (2.30).

$$\sigma_{IR} = \frac{\sqrt{\Delta I^2}}{2\gamma I_0} \sqrt{\frac{1}{\sin^2\alpha_t} + \frac{3}{(1 - \cos\alpha_t)^2}} \qquad (2.30)$$

Thereby, the proportionality $\sigma_{IR} \propto \frac{\Delta I}{2\gamma I_0}$ is in general valid for the phase shifting methods.

Reconstruction of the Object Wave Amplitude

In order to determine the amplitude of the object wave in a temporal phase shifting digital holographic interferometer, the modulation $\gamma(x_n, y_m)$ and $I_0(x_n, y_m)$ is required. For the three step algorithm (2.25) these are given by [6]:

$$\gamma(x_n, y_m) = \frac{\sqrt{(2I_j - I_{j-1} - I_{j+1})^2 + (I_{j-1} - I_{j+1})^2 \tan^2 \frac{\alpha_t}{2}}}{I_{j-1} + I_{j+1} - 2I_j \cos \alpha_t}, \qquad (2.31)$$

$$I_0(x_n, y_m) = I_R(x_n, y_m) + I_O(x_n, y_m) = \frac{2 \cos \alpha_t I_j - I_{j-1} - I_{j+1}}{2 \cos \alpha_t - 2}. \qquad (2.32)$$

Assuming a plane reference wave with constant intensity $(I_R(x_n, y_m) = I_R)$, (2.24) this leads to the following amplitude distribution:

$$\sqrt{I_0(x_n, y_m)I_R} = \gamma(x_n, y_m) \cdot \frac{1}{2} \cdot (I_R(x_n, y_m) + I_O(x_n, y_m)) \qquad (2.33)$$
$$\propto |E_O(x_n, y_m)|.$$

In case of a known modulation γ and phase ϕ_0 the complex object wave in the hologram plane (CCD-sensor) can be approximated with one another by the following equation [6]:

$$E_O(x_n, y_m) \propto I_0(x_n, y_m)\gamma(x_n, y_m)e^{i\phi_0(x_n, y_m)}. \qquad (2.34)$$

---

[2]In literature, a phase step is usually expressed in radians. Since the degree unit is more vivid, it is preferred in this work.

2.2.2   Spatial Phase Shifting

In spatial phase shifting techniques the hologram is recorded in off-axis geometry by a tilt between object and reference wave, in order to overlay a system of (carrier) fringes onto the hologram, and therefore creating a phase gradient (α_x, α_y) on purpose in (2.29).

For the reconstruction of the phase and amplitude of the object wave, the corresponding grey-scale values of neighbouring pixels, in the direction perpendicular to the phase gradient, can be inserted into the interferogram equation (2.22) in analogy to temporal phase shifting. This results in a determined and thus solvable system of equations with three different unknown variables (I_0, γ, φ).

Another procedure which has proven to be particularly suited for this problem will be described in the following. It represents a generalised solution to spatial phase shifting problems [25, 26]. It is a spatial phase shifting technique for the reconstruction of the complex object wave (phase and amplitude) specially adapted for a microscopy setup for digital "off-axis" holography. The main advantage of this - sometimes denoted "non-diffractive" - reconstruction method (NDRM) is, that the zero diffraction order and the twin image are excluded from the reconstruction.

The algorithm is based on the assumption, that only the relative phase difference $\varphi(\vec{r})$ between object wave $E_O(\vec{r})$ and reference wave $E_R(\vec{r}) = E_{OR}(\vec{r}) \exp(i\varphi(\vec{r}))$ cause high spatial frequency changes in the intensity distribution $I_H(\vec{r})$ of the digital hologram, so that the object wave in the vicinity of a pixel $\vec{r} = (x, y)$ of M pixels $(i = 1, \ldots M)$ can be assumed as slowly varying and thus approximately constant [27]. Under this assumption the complex object wave $E_O(\vec{r})$ at the point $\vec{r}$ in the hologram plane can be calculated by solving a nonlinear system of M equations, which correspond to the interferogram equations at the position $\vec{r} + \vec{r}_i$.

$$I_H(\vec{r} + \vec{r}_i) = |E_O(\vec{r}) + E_{OR}(\vec{r}) \exp(i\varphi(\vec{r} + \vec{r}_i))|^2 \qquad (2.35)$$
$$\text{with} \qquad E_O(\vec{r}) \approx E_O(\vec{r} + \vec{r}_i)$$
$$\text{and} \qquad E_{OR}(\vec{r}) \approx E_{OR}(\vec{r} + \vec{r}_i).$$
$$I_i = |E_O + E_{OR} \exp(i\varphi)|^2 \qquad (2.36)$$

In (2.36) and in the following the spatial dependence will be omitted to improve legibility. The resulting equation system is solved with the method of least squares. The solution for (2.36) is given in (2.37). Complex conjugated variables are indicated by a "*".

$$E_O = \frac{1}{E_{OR}} \frac{(b - va)(1 - v^*v) - (c - v^*a)(w - vv)}{(1 - vv^*)(1 - v^*v) - (w - vv^*)(w - vv)} \qquad \text{mit:} \qquad (2.37)$$

$$v = 1/M \sum_i V_i, \; w = 1/M \sum_i V_i^2, \; a = 1/M \sum_i I_i, \; b = 1/M \sum_i V_i I_i, \; c = 1/M \sum_i V_i^* I_i,$$

$$V_i = E_{Ri}^* / E_{0R} = \exp(-i\varphi) \tag{2.38}$$

The calculations assume the reference wave's amplitude to be constant ($E_{0R}(x, y) \equiv E_{0R}$). The complex object wave is determined pixel wise by calculating the sums v, w, a, b and c in (2.37). For the reconstruction of E_O with (2.37) the intensity distribution $I_H(x_n, y_m)$ and the phase distribution $\varphi(x_n, y_m)$ must be known. The mathematical model for the phase distribution is given in (2.39) and (2.40):

$$\varphi(x_n, y_m) = 2\pi(K_x x_n^2 + K_y y_m^2 + L_x x_n + L_y y_m) \tag{2.39}$$

$$K_x = \frac{\Delta x^2}{2\pi} \varphi_x = \frac{\Delta x^2}{\lambda^2 + 2\lambda d} \quad \text{and} \quad K_y = \frac{\Delta y^2}{2\pi} \varphi_y = \frac{\Delta y^2}{\lambda^2 + 2\lambda d} \tag{2.40}$$

The quadratic terms in (2.40) describe a phase gradient which is formed by the superposition of a plane reference wave with a spherical object wave, whose source has the distance d from the hologram plane. This is the case in special setups such as microscopic arrangements [26]. In case of two parallel waves the quadratic terms vanish. The linear terms describe the constant phase gradient which is given by the angle between object and reference wave (α_x, α_y), the pixel pitch of the scanning sensor (Δx, Δy) and the wavelength λ.

$$L_x = \frac{\Delta x}{2\pi} \varphi_x = \frac{\beta_x \Delta x}{\lambda} \quad \text{and} \quad L_y = \frac{\Delta y}{2\pi} \varphi_y = \frac{\beta_y \Delta y}{\lambda} \tag{2.41}$$

One of the disadvantages of spatial phase shifting is the averaging effect in the calculation of E_O that arises in addition to the pixel integration. This usually causes a slight reduction of the effective lateral resolution. In microscopy this can be compensated partially by oversampling with an imaging sensor. Since only one recorded hologram is needed, the advantages are a low susceptibility towards oscillations and an easily realizable setup.

2.3   Numeric Propagation of Complex Object Waves

Using numerical propagation, the object wave can be reconstructed in different planes and thereby focused later. In digital holographic microscopy this method permits a subsequent (re)focusing of, for instance, several cells in different axial planes and therefore a simultaneous sharp imaging of multiple cells within the depth of field of a microscope objective [28, 29].

To numerically propagate the reconstructed object wave from the hologram plane (x, y) into an image plane (x', y') with a distance Δz between them (see Fig. 2.4) the Huygens–Fresnel principle presented in Sect. 2.1.5 is being used. The integral (2.20) in this work is being calculated with the convolution method (CVM) [6].

$$E_{\Delta z}(x', y') = \int \int_{-\infty}^{\infty} E_0(x, y)h(x' - x, y' - y)\mathrm{d}x\,\mathrm{d}y \qquad (2.42)$$
$$= E_0(x', y') * h(x', y')$$

with the unit pulse response:

$$h(x', y') = \frac{e^{i\frac{2\pi\Delta z}{\lambda}}}{i\lambda\Delta z} \exp\left(i\frac{\pi}{\lambda\Delta z}\left(x'^2 + y'^2\right)\right) \qquad (2.43)$$

where $*$ is the convolution operation. Applying the Fourier transform (FT) and the convolution theorem on (2.42) yields:

$$\mathrm{FT}\left(E_{\Delta z}(x', y')\right) = \mathrm{FT}\left(E_0(x', y') * h(x', y')\right) \qquad (2.44)$$
$$= \mathrm{FT}\left(E_0(x', y')\right) \cdot \mathrm{FT}\left(h(x', y')\right)$$

After application of the inverse Fourier transform and transformation into a discrete representation with the coordinates (x_n, y_m) and assuming a raster image field of size $N_x \times N_y$ with pixel sizes of Δx and Δy, (2.44) yields:

$$E_{\Delta z}(x_n, y_m) = \frac{1}{i\lambda\Delta z} e^{\frac{i2\pi\Delta z}{\lambda}} \qquad (2.45)$$
$$\mathrm{FFT}^{-1}\left\{\mathrm{FFT}\left\{E_0(x_n, y_m)\right\} \exp\left(i\pi\lambda\Delta z\left(\frac{n^2}{\Delta x^2 (N_x)^2} + \frac{m^2}{\Delta y^2 (N_y)^2}\right)\right)\right\}$$

FFT represents the Fast Fourier Transform. The quadratic phase factor in (2.45) fulfils the sampling theorem under the condition

$$|\Delta z| \leq \frac{\Delta x^2 N_x}{\lambda}, \qquad |\Delta z| \leq \frac{\Delta y^2 N_y}{\lambda}. \qquad (2.46)$$

For big propagation distances the data field has to be filled with zeros at the boundaries in order to avoid aliasing, which is also known as zero padding. In contrast to other numerical evaluations of (2.17) like the Discrete Fresnel Transformation (DFT) the sampling interval remains constant during the transformations based on the CVM ($\Delta x' = \Delta x$ bzw. $\Delta y' = \Delta y$) [6] and the image size does not change during propagation.

2.3.1 Digital Holographic Microscopy

Digital holographic microscopy (DHM) employs digital holography to allow quantitative phase contrast images. In microscopy, Zernicke- or Nomarski methods are commonly used which correspond to phase contrast microscopy and differential

Fig. 2.5 For the imaging with digital holographic microscopy a laser beam is splitted into object and reference wave. The object wave passes through the object plane and is superimposed with the reference wave in a tilt angle ("off-axis") so that the interference can be captured with a CCD camera

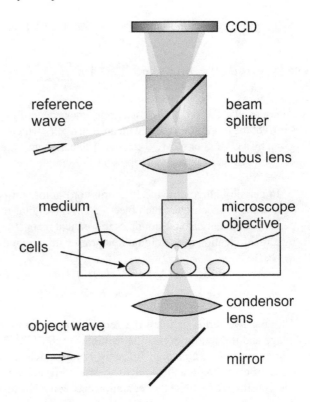

interference contrast. For a simple image of a sample with low absorption power these methods are appropriate. However, to obtain a quantitative result with these methods alone is often not possible. The image artefacts of the Zernicke method (Halo-, Shading-Off- and Lens-effects) prohibit a quantitative result. The Nomarski method can produce exact quantitative measurements, but only with a high technical effort. This can be realised - in analogy to temporal phase stepping techniques - by measuring 4 different interferograms that are temporally phase shifted by 90 which usually demands an axial translation of the object plane precise to sub wavelength lengths [21]. Digital holographic microscopy also facilitates an exact quantitative measurement with low requirements concerning the optical platform. The quantitative results can be used for measurements on reflective as well as transparent specimens in $2\,1/2$ dimensional and pseudo-3D representations [4, pp. 58ff.].

For quantitative digital holography, a laser beam is splitted between object and reference beam (see Fig. 2.5). The object wave illuminates the object plane and is phase shifted as a function of the refractive index distribution of the sample. Object and reference wave are subsequently superimposed with another beam splitter and brought to interference. The resulting intensity pattern of the hologram $I_H(x, y)$ (see also Fig. 2.6) can be then described by (2.22). In digital holography this pattern is not recorded by a special film, but digitised by a CCD camera. The digitised hologram

$I_H(k, l)$ results from the spatial sampling of the CCD matrix chip:

$$I_H(k, l) = I_H(x, y)\text{rect}\left(\frac{x}{L}, \frac{y}{L}\right) \times \sum_k^N \sum_l^N \delta(x - k\Delta x, y - l\Delta y), \quad (2.47)$$

Alternatively, the preferably quadratic image recording chip or region of interest (ROI) of the hologram matrix with an area $L \times L$, which consists of $N \times N$ pixels, is given by a CMOS or sCMOS camera. The parameters k and l are integers ($-N/2 \leq k, l \leq N/2$) and Δx and Δy represent the sample intervals in the hologram plane; $\Delta x = \Delta y = L/N$.

In classic holography, the (transmission) hologram is illuminated with a reference wave to reconstruct the object (2.7). The digital reconstruction in contrast is performed numerical and simulates this optical reconstruction. This can be achieved by a multiplication of the digital hologram with a digital calculated reference wave $R_D(k, l)$:

$$E_D(k\Delta x, l\Delta y) = R_D(k, l)I_H(k, l) = R_D |R|^2 + R_D |O|^2 \\ + R_D R^* O + R_D R O^*. \quad (2.48)$$

The first two terms here are the zero diffraction order, the third term is the twin image and the fourth term is the real image (cf. (2.7)). The tilt between object and reference wave results in an "off-axis" setup which allows for a calculation of the three different holographic images shown in Fig. 2.2 which are then prevented from superposing. To increase the separation between real and twin image, the tilt angle of the two waves should be sufficiently high. The maximum angle however is limited by the resolving capability of the CCD sensor, since the spatial frequency of the interference pattern needs to be small enough to be fully resolved by the imaging chip [5, p. 37ff.], but sufficiently low to fulfil the Nyquist theorem.

For an error free reconstruction it is beneficial when the digital calculated reference wave R_d is an exact emulation of the reference wave R which was used in the hologram acquisition. If this is the case, the product $R_D R^*$ in (2.48) becomes real ($RR^* = |R|^2$). This enables the reconstruction of the phase of the twin image O. An additional possibility would be given by using a digital reference wave, which is the complex conjugate of R. In that case the real image O^* can be calculated. If the reference wave is assumed to be a plane wave with wavelength λ, then R_D yields:

$$R_D(k, l) = A_R \exp\left[i\frac{2\pi}{\lambda}(k_x k\Delta x + k_y l\Delta y)\right], \quad (2.49)$$

with the two components of the wave vector k_x and k_y and the amplitude A_R. Equations (2.48) and (2.49) yield the complex, digital transmitted wave front at the spot in the hologram plane. To obtain the wave front of a different plane, e.g. the object plane, a numerical propagation is required.

In general the propagation of $E_D(k\Delta x, l\Delta y)$ is realised by a numerical calculation of the scalar diffraction integral in the Fresnel approximation similar to Sect. 2.1.5.

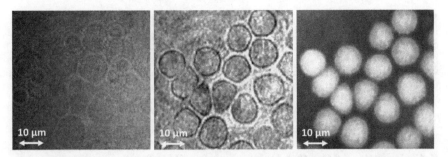

Fig. 2.6 Exemplary digital holographic microscope image of erythrozyts with the digitised hologram (left), the numerical reconstructed amplitude (middle) and phase image (right)

Here, the reconstructed wave front $E_D(m\Delta\xi, n\Delta\eta)$ in an observation plane $O_{\xi\eta}$ with an axial distance Δz to the hologram plane is calculated with the discretised form of the Fresnel integral [30]:

$$E_D(m\Delta\xi, n\Delta\eta) = A \exp\left[\frac{i\pi}{\lambda\Delta z}(m^2\Delta\xi^2, n^2\Delta\eta^2)\right]$$
$$\times FFT\left\{R_D(k,l)I_H(k,l)\exp\left[\frac{i\pi}{\lambda\Delta z}(k^2\Delta x^2, l^2\Delta y^2)\right]\right\}_{m,n},$$
(2.50)

where m and n are integer (analogue to k and l), FFT represents the "Fast Fourier transform"-Operation and $A = \exp(i2\pi\Delta z/\lambda)/(i\lambda d)$. The parameters $\Delta\xi$ and $\Delta\eta$ are along the sample intervals in the observation plane and denote the lateral resolution of the reconstructed image as a function of the size L of the CCD and the distance Δz:

$$\Delta\xi = \Delta\eta = \lambda\Delta z/L. \tag{2.51}$$

The reconstructed complex wavefront from (2.50) is an array of complex numbers. The amplitude and phase representation are then retrieved from an calculation of the absolute value and phase:

$$\text{absolute value} = \sqrt{Re(E_D)^2 + Im(E_D)^2}$$
$$\text{Phase} = \arctan\left(\frac{Im(E_D)}{Re(E_D)}\right) \tag{2.52}$$

The calculated phase values can only have values in the interval $[-\pi, \pi]$. In case the sample has height differences exceeding λ the phase image will show phase jumps. Employing phase-unwrapping methods these images can be transformed to continuous phase values [31] that can be displayed in a 3D representation (see also Fig. 2.7).

Fig. 2.7 Pseudo 3D representation of the quantitative results of the reconstructed phase image shown in Fig. 2.6. The cell thickness is $\approx 1, 4\,\mu m$

2.4   Benefits of Partial Coherence for DHM

2.4.1   Spatial Frequency Filtering

The objective of this section is to investigate the relationship between spatial coherence and digital holographic reconstruction. The theoretical analysis outlines the relationship between coherence and digital holographic reconstruction [32].

In order to have a uniform influence on the field of view, it is significant for the spatial coherence of the illumination to be the same for every point of the field of view (FOV). Therefore, it is assumed that the spatial coherence of the light distribution illuminating the sample is represented by a stationary function $\Gamma(x_1 - x_2, y_1 - y_2)$, where (x_1, y_1) are spatial coordinates in the FOV of the microscope in the object channel (Fig. 2.5).

Performing a double Fourier transform of $\Gamma(x_1 - x_2, y_1 - y_2)$ on the spatial variables yields according to [32]

$$F_{x_1,y_1} F^{-1}_{x_2,y_2} \Gamma(x_1 - x_2, y_1 - y_2) = \gamma(u_2, \upsilon_2)\delta(u_1 - u_2, \upsilon_1 - \upsilon_2) \ , \qquad (2.53)$$

where (u_k, υ_k) are the spatial frequencies associated with (x_k, y_k).

In (2.53), the presence of the Dirac function δ indicates that the Fourier transformed $\Gamma(x_1 - x_2, y_1 - y_2)$ can be seen as an incoherent light distribution modulated by the function $\gamma(u_2, \upsilon_2)$. Therefore, regardless of the exact way of how a partially spatial coherent illumination is obtained, it can be considered that it is built from an apertured spatial coherent source placed in the front focal plane of a lens. As it corresponds exactly to the way a source is prepared in the microscope working with a laser and (an optional) rotating ground glass, this lens is denoted L2, with the focal length f_2 in Fig. 2.8. The transparency in a plane UP is denoted with $t(x, y)$. It is assumed that the plane FP, the back focal plane of L2, is imaged by the microscope lens on the CCD camera.

In the following, the diffraction and coherence effects on $t(x, y)$ in the plane FP are investigated, separated from UP by a distance d. It is advantageous that the secondary source in the plane of the apertured incoherent source (AIS) is completely incoherent. Therefore, the computation of the partial coherence behaviour can be derived as follows.

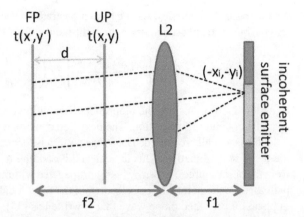

Fig. 2.8 Principle of a partly coherent source given by a surface emitter (e.g. LED)

First an amplitude point source in the *AIS* plane is regarded. Then the intensity distribution in the output plane due to this point source can be computed. Finally, one can integrate the result over the source area in the rotating ground glass plane.

As next, an amplitude point source $\hat{\delta}(x + x_i, y + y_i)$ located in the AIS plane is considered. The amplitude distribution that is illuminating the transmission plane $t(x, y)$ is a tilted plane wave in such a way that the amplitude distribution emerging out of $t(x, y)$ is expressed by

$$u_{\mathrm{UP}}(x, y) = Ct(x, y) \exp\left(j2\pi \frac{xx_i + yy_i}{\lambda f_2}\right) , \tag{2.54}$$

where C is a constant. To determine the amplitude distribution in FP, the free space propagation operator [33] on the amplitude distribution given by (2.54) is utilised:

$$
\begin{aligned}
u_{\mathrm{FP}}(x', y') = {}& C' \exp\left\{j2\pi \frac{x_i x' + y_i y'}{\lambda f_2}\right\} \\
& \times \exp\left\{-j\frac{2\pi d}{\lambda f_2^2}(x_i^2 + y_i^2)\right\} T_f\left(x' - \frac{d}{f_2}x_i, y' - \frac{d}{f_2}y_i\right) ,
\end{aligned}
\tag{2.55}
$$

with

$$
\begin{aligned}
T_f&\left(x' - \frac{d}{f_2}x_i, \ y' - \frac{d}{f_2}y_i\right) \\
&= \int\int dv_x dv_y \exp\left\{j2\pi\left[v_x(x' - \frac{d}{f_2}x_i) + v_y(y' - \frac{d}{f_2}y_i)\right]\right\} \\
&\quad \times \exp\left\{-j\frac{k\lambda^2 d}{2}\left[v_x^2 + v_y^2\right]\right\} T(v_x, v_y) ,
\end{aligned}
\tag{2.56}
$$

where C' is a constant including terms that are not playing a significant role, $T(v_x, v_y)$ is the continuous spatial Fourier transform of $t(x, y)$ defined by (2.53), and (v_x, v_y)

are the spatial frequencies. As can be demonstrated on actual examples, it is assumed
that the quadratic phase factor on the right-hand side of (2.55) can be neglected [33].

$$u_{FP}(x', y') = C' \exp\left\{j2\pi\frac{x_i x' + y_i y'}{\lambda f_2}\right\} T_f\left(x' - \frac{d}{f_2}x_i, y' - \frac{d}{f_2}y_i\right) . \quad (2.57)$$

Equation (2.56) describes the amplitude distribution obtained in the FP plane for a
spatially coherent illumination. The next step consists in evaluating the interference
field that is actually recorded. The fringe pattern is recorded by the image sensor after
the beam recombination. This is achieved by adding to (2.57) the beam originated
from the point source $\delta(x + x_i, y + y_i)$ propagated without transparency in the optical
path. In case the point source is located in the front focal plane of lens L2, the beam
in plane FP is a tilted plane wave that is written as [33]

$$u_{REF}(x', y') = C'' \exp\left\{j2\pi\frac{x_i x' + y_i y'}{\lambda f_2}\right\} \quad (2.58)$$

where C'' is a constant.

Equations (2.57) and (2.58) are added and the square modulus of the results is
computed to achieve the light intensity

$$i_{FP}(x', y')|u_{REF}(x', y') + u_{FP}(x', y')|^2 = |C''|^2 + |C'T_f|^2 + AT_f + A^*T_f^* \quad (2.59)$$

where $A = C''^* C'$, and the explicit spatial dependency of T_f was cancelled for the
sake of simplicity. The two terms linear with T_f and T_f^*, at the right-hand side of
(2.59), are the holographic signals. As a phase stepping technique or the Fourier
transform method is applied, the phase and amplitude modulus of T_f are the significant information extracted from (2.59). Considering now the effect of the source
extension by integration over the source domain. The signal $v(x', y')$ that will actually
be detected can be expressed by [32]

$$v(x', y', d) = A \int\int Ip^2(x_i, y_i)T_f\left(x' - \frac{d}{f_2}x_i, y' - \frac{d}{f_2}y_i\right) dx_i dy_i . \quad (2.60)$$

Equation (2.60) shows that the signal $v(x', y', d)$ is the correlation product between
T_f and the source function $p^2(x_i, y_i)$. By performing the change of the integration variables $x_i' = x_i d/f_2$, $y_i' = y_i d/f_2$ and by invoking the convolution theorem, one obtains

$$V(v_x, v_y, d) = \left[F^{+1}_{(C)v_x,v_y}T_f(x', y')\right]S\left(\frac{v_x d}{f_2}, \frac{v_y d}{f_2}\right) \quad (2.61)$$

where V, $[F^{+1}_{(C)x',y'}T_f(x', y')]$, and S are the Fourier transform of v, T_f and p^2. Equation (2.61) basically shows that the Fourier transform of the detected signal in partially coherent illumination corresponds to the Fourier transform of the signal with a

coherent illumination filtered by a scaled Fourier transformation of the source. The scaling factor d/f_2 means that the filtering process increases with d. As $v(x', y', d)$ is the amplitude that is used to perform the digital holographic reconstruction, a loss of resolution of the reconstructed focus image occurs due to the partial spatial coherent nature of the illumination.

Equation (2.61) allows to set accurately the partial coherence state of the source with respect to the requested resolution, refocusing distance, and the location of the optical defects that have to be rejected. As an example in biology, it often happens that the selection of the experimental sample cell cannot be made on the only criteria of optical quality but has also to take into account the biocompatibility. Therefore, the experimental cells are often low-cost plastic containers with optical defects. In this situation, the partial coherence of the source can be matched to refocus the sample by digital holography while keeping the container windows at distances where the spatial filtering described above is efficient to reduce the defect influence.

With (2.61), it is expected to have an increasing resolution loss with the distance between the refocused plane and the best focus plane. However, the loss of resolution is controlled by adjusting the spatial partial coherence of the source and, in this way, can be kept smaller than a limit defined by the user. It has also to be emphasised that the reduction in spatial coherence is a way to increase the visibility of the refocused plane by reducing the influence of the out of focus planes. This aspect is particularly useful when the sample highly scatters the light. The adjustment capability of the spatial coherence represents then a very useful tool for tuning the image visibility and the depth of reconstruction with respect to the sample characteristics.

2.4.2 Straylight and Multiple Reflection Removal

Partial coherent illumination opens up prospect to avoid multiple reflections that can occur with coherent illumination [34]. This is obvious when applying an LED illumination due to its small temporal coherence. In this case, it is assumed that a reflection introduces an increase of the optical path d. If the distance d introduces a significant decorrelation of the speckle pattern, the contrast of the interference fringe pattern between the reflected beam and the direct beam is reduced. Next, the geometry of Fig. 2.8 is regarded. As in the image plane, the instantaneous amplitude field can be expressed by the product of a random phase function r by an amplitude modulation p. The function r is defined by [6]

$$r(x, y) = \exp\{j\phi(x, y)\} \tag{2.62}$$

where $\phi(x, y)$ is a random function with a constant probability density on the interval $[O, 2\pi]$. It is also assumed that $\phi(x, y)$ is very rapidly varying in such a way that

$$\langle r^*(x', y')r(x, y)\rangle = \delta(x' - x, y' - y) , \tag{2.63}$$

where $\langle\rangle$ denotes the ensemble average operation.

Therefore, the instantaneous speckle amplitude field in the object plane is expressed by [6]

$$s(x, y) = \exp\{j2kf_2\} \, (P \otimes R) \left(\frac{x}{\lambda f_2}, \frac{y}{\lambda f_2} \right) , \tag{2.64}$$

where R and P are the Fourier transformations of, respectively, p and r. It is considered that in the optical path a double reflection on a window is present that is introducing an additional optical path d. The double reflected beam is here denoted with $s'(x', y')$. It can be seen as the $s(x, y)$ beam propagated by a distance d:

$$s'(x', y') = A \exp(jkd) \left[F_{(C)x', y'}^{-1} \exp(-\frac{jkd\lambda^2}{2}(v_x^2 + v_y^2)) \left[F_{(Cv_x, v_y)}^{+1} s(x, y) \right] \right] , \tag{2.65}$$

where A defines the strength of the double reflection. Inserting (2.64) in (2.65), we obtain

$$s'(x', y') = A \exp(ikd)[\, F_{(C)x', y'}^{-1} \exp\left(-\frac{jkd\lambda^2}{2}(v_x^2 + v_y^2) \right)$$
$$\times \ r(-v_x\lambda f_2, -v_y\lambda f_2) \, p(-v_x\lambda f_2, -v_y\lambda f_2)\,] , \tag{2.66}$$

If the quadratic phase factor is slowly varying over the area where the p-function in (2.66) is significantly different to zero, $s'(x', y')$ is very similar to $s(x, y)$ and they mutually interfere to result in a disturbing fringe pattern. On the contrary, when the quadratic phase factor is rapidly varying on the significant area defined by the width of p, $s'(x', y')$ and $s(x, y)$ are uncorrelated speckle fields and the interference modulation visibility is reduced when the ground glass is moving. Assuming that the laser beam incident on the ground glass has a Gaussian shape, p is expressed by

$$p(x, y) \propto \exp\left\{ -\frac{x^2 + y^2}{\omega^2} \right\} , \tag{2.67}$$

where ω is the width of the beam. Using (2.67), the width of $p(-v_x\lambda f_2, -v_y\lambda f_2)$ in (2.66) is equal to $\omega^2/\lambda^2 f_2^2$. It can be assumed that one obtains a large speckle decorrelation when the quadratic phase factor in the exponential of (2.66) is larger than π, and when $(v_x^2 + v_y^2)$ is equal to the width of $p(-v_x\lambda f_2, -v_y\lambda f_2)$. Therefore, a speckle decorrelation between $s'(x', y')$ and $s(x, y)$ is achieved when

$$\frac{d\omega^2}{\lambda f_2^2} >> 1 . \tag{2.68}$$

As the width ω is adjusted by changing the distance between a lens and a rotating ground glass (RGG), multiple reflection artefacts are removed by a difference path distance d appropriately reduced.

2.5 Types of Spatial Light Modulators

2.5.1 Different Methods of Addressing

Spatial light modulators (SLM) are devices that allow the user to modulate the amplitude and/or phase of a light beam. The addressing of the SLM can be performed optically or electronically.

An optically addressed SLM - also called Pockels Readout Optical Modulator (PROM) - consists of suitable crystals (e.g. bismuth silicon oxide) and is based on the Pockels effect. The birefringent crystal is located between two transparent electrodes that are supplied with a voltage. There is a dichroic mirror which transmits blue and reflects red light (see Fig. 2.9) between two electrodes. In addition, the applied crystal material is photoconductive at blue wavelengths so that a charge and therefore voltage distribution proportional to the light intensity can be induced in the crystal with a blue writing beam. This causes a local change of the refractive index. The written information is then read using a beam at red wavelengths. This modulates phase and/or amplitude of the red read beam. A disadvantage of this method is the high operating voltage of the electrodes which is in the range of several kV [35, pp. 731f.]. Electrically addressed SLMs consist of a pixel array wherein each pixel is addressed individually via application of a voltage - similar to conventional electronic displays. Just like the latter, these SLMs are usually controlled with a video card output and therefore a VGA, DVI or HDMI port on a PC. Here, it can be distinguished between Digital Micromirror Devices (DMDs) and SLMs based on Liquid Crystal (on Silicon) (LCD/LCoS), which are presented in the following section [36, pp. 202ff.].

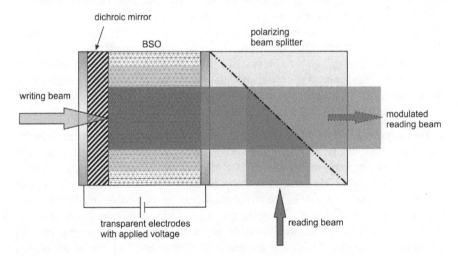

Fig. 2.9 A writing beam induces a voltage distribution into a photoconductive crystal located between two transparent electrodes which apply a high voltage. A reading beam of a different wavelength is passed through this crystal and gets modulated by the local changes of the refractive index in the crystal

Fig. 2.10 The mirrors of a DMD can typically be tilted in a range of ±13° [37] with a frequency of ≈[5, 32] kHz and therefore modulate the intensity of a beam pixel by pixel

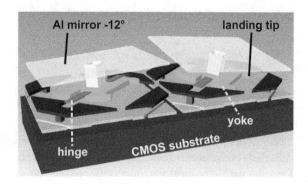

2.5.2   Digital Micromirror Devices and Liquid Crystal SLMs

Digital micromirror device (DMD): DMDs are (among others, developed and distributed by the Fraunhofer IPMS and Texas Instruments) pixelated components in which each pixel consists of an aluminium micro-mirror and additional elements for addressing and movement. The modulation of the amplitude or phase of a reflected light beam on a DMD is based on a translation of individual mirrors (Fig. 2.10).

An amplitude modulation is realised by tiltable mirrors. Each mirror is tilt-mounted in its centre and by applying a voltage it can be tilted in a typical range of [−13°, +13°]. This enables to change between the states "on" and "off". If the mirror is not tilted, the light beam is reflected (on). By tilting the mirror, the light beam is often - depending on the application - deflected onto an absorber, so that the pixel is not reflecting any light into the beam path (off). Since the mirrors can change their state with a frequency in the range of of ≈[5, 32] kHz, it is possible to realise up to 1024 gradations of intensity by changing the time difference between the "on" and "off" state which corresponds to a fluctuating state. These components are mainly used in projectors [38, pp. 8ff.] [39]. While an intensity modulation is achieved by tiltable mirrors, a phase modulation can be generated by translation of the single mirrors. This changes the geometric path length of the reflected light and therefore enables the user to modulate the phase of the beam [40].

Liquid crystal SLM: SLMs based on liquid crystals are pixelated semiconductor elements which can be distinguished according to their structure (transmissive or reflective) and its main modulation (phase-only, amplitude-only, etc.). A transmissive liquid crystal SLMs consist of a light-transmissive LCD that is composed of several layers (see Fig. 2.11). The cover glass is used as a stable mechanical support structure and to protect the screen. The subsequent electrically conductive and light-transmissive ITO (Indium Tin Oxide) layer functions as a counter electrode to the address electrodes. After an alignment layer which serves for spatial orientation of the LC molecules follows the liquid crystal layer. The other configuration is mirrored to the previous, with the difference that the layer of the addressing electrode is pixelated.

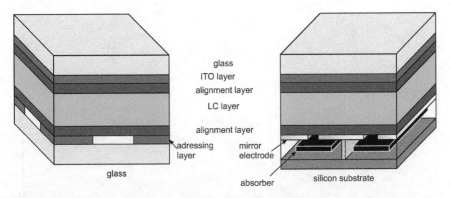

Fig. 2.11 Schematic structure of an LCD (left) and an LCoS screen (right)

The reflective liquid crystal modulators are LCoS screens which have a similar layout to the LCDs. However, the addressing layer consists of reflective mirror electrodes which are located on a silicon substrate. Thus, these aluminium-coated electrodes represent both the electronic and the optical interface to the LC layer. Absorbers collect rays that pass through the small gap between two mirror electrodes. The advantage of a reflective SLM is the substantially higher fill factor, as the electronics for driving the pixels is located behind the reflective layer and not between the pixels [41].

The principle of modulation is determined by the orientation of the alignment layers in combination with the polarisation of the light. By applying a voltage to the LC layer, the orientation of the crystals leads to a change in the refractive index proportional to the strength of the applied voltage. Due to the birefringent[3] characteristics of the LC, light experiences different refractive indices depending on the LC-molecular orientation relative to the electric field component and the direction of propagation of the light. The two refractive indices (ordinary n_o and extraordinary n_{ao}), which act perpendicular to each other, lead to a wavelength-dependent delay γ of the passing light λ in a LC layer of thickness d:

$$\Gamma = \frac{2\pi d (n_{ao} - n_o)}{\lambda}. \tag{2.69}$$

With a suitable orientation of the molecules it is possible to rotate the polarisation direction of linearly polarised light. The rotation angle can be adjusted by the strength of the applied voltage. Thus amplitude modulation can be realised in combination with linear polarising filters (see also Fig. 2.12). If the light's polarisation is turned so that it hits the polarising filter in the longitudinal direction, it can pass completely. If it is rotated 90 it is fully absorbed [43]. Phase modulation is achieved with the propagation delay from (2.69). By increasing the refractive index the propagation speed in the material decreases and therefore the light takes longer to pass

---

[3]For a more detailed explanation, please refer to the corresponding literature [42, pp. 547ff.].

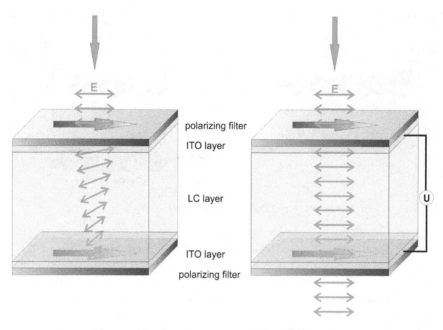

Fig. 2.12 Twisted nematic, "normally black" liquid crystal cell without (left) and with applied voltage (right). The double arrows represent the direction of the electric field vector and the arrows on the polarising filters visualize the polarisation direction

the material. Early "phase-only" LCoS screens had slow switching of the refractive index which led to inadequate results, especially in dynamic applications. The slow behaviour arises from the relaxation time of the LC molecules, that change their orientation relatively slow with a change in voltage. The latest generation of screens circumvents this problem with LC molecules which do not orientate themself into the original state, but are actively switched back using a special high frequency (40kHz) addressing technique [44]. The SLMs used in this work are exclusively liquid crystal modulators with a "phase-only" modulation (see also characterisation in Sect. 4.5). In a phase-only SLM, the phase of the reflected beam is changed only by utilizing an outer voltage altering the orientation of LC molecules that also depends on parameters like viscosity and elasticity of the medium [45]. When voltage is modulated sufficiently quick, the LC molecules sense solely their average value. Consequently, modulations slower than the relaxation time of the LC are rather followed by the molecules which usually induces a time-dependent phase flicker [45, 46].

2.5.3   Light Modulators as Holographic Elements

In order to use light modulators as holographic elements, special properties are needed for the reconstruction of computer-generated holograms. Significant assess-

Table 2.1 Diffraction efficiency of different hologram structures

Diffractive structure	Maximum diffraction efficiency (%)
Continuous sine amplitude grating	6
Continuous sine phase grating	34
16 stage sine phase grating	33.6
8 stage sine phase grating	33.2
4 stage sine phase grating	27.5
Binary amplitude grating	10
Binary phase grating	41
Blazed phase grating	100

ment parameters for the quality of a reconstructed hologram are the diffraction efficiency and the spatial bandwidth product, which are treated in the following:

Diffraction efficiency: The diffraction efficiency η_{diff} is derived from the ratio of the intensity in the first diffraction order I_1 and the total intensity I_{tot}:

$$\eta_{\text{diff}} = \frac{I_1}{I_{\text{tot}}}. \tag{2.70}$$

The diffraction efficiency highly depends on the nature of the hologram. Amplitude modulated holograms have a much lower diffraction efficiency than phase modulated holograms, as the intensity is reduced to modulate the light. Table 2.1 gives an overview over the diffraction efficiency of different simple amplitude and phase modulated holograms [47, pp. 44f.].

Local bandwidth product: The local bandwidth product describes the amount of information a signal contains. It is defined as the product of the expansion of a signal in position-space L_x, L_y and its expansion in the frequency domain f_x, f_y:

$$M = L_x L_y f_x f_y. \tag{2.71}$$

The content of holograms can be compared with the number of addressable pixels on the SLM, thus the spatial bandwidth product is the product of the resolution and size of the SLM.

In addition, image artefacts like strong higher orders of diffraction are enhanced by a non-linear characteristic curve or low fill factors. The diffraction efficiency of an SLM with absorbing areas between the pixels is given by:

$$\eta_{\text{SLM}} = \bar{R} \cdot (1 - \bar{A}) \cdot FF^2 \cdot \eta_H \cdot \eta_{\text{on}}. \tag{2.72}$$

The parameter $\bar{R}$ represents the reflectivity of the mirror surface, $\bar{A}$ the absorption effects by internal reflections or unwanted amplitude modulation, and FF the fill factor of the screen. The efficiency of the hologram is depicted by η_H in the equation and η_{on} corresponds to the fraction of usable time which has an effect on the time-dependent modulation behaviour during switching (e.g. in DMDs) [48, pp. 43ff.]. In summary, a light modulator should have the following properties in order to ensure a high reconstruction quality:

- high diffraction efficiency
- high fill factor
- linear phase shift between 0 and 2π
- sufficiently high local maximum-bandwidth product

LCoS displays combine these characteristics best. They are available as a "phase-only" variant where a linear phase shift of 0 to 2π at very low amplitude modulation is possible. In addition, they have a high fill factor (>90%) as the conducting paths and electronics are behind the mirror surface and not between the pixels as in the LCD version. Because of the relatively high resolution of typically up to 1920×1080 pixels, they also offer a sufficient maximal local bandwidth product.

The development of phase shifting DMDs is promising for usage as holographic elements, since they have a linear phase shift from 0 to over 2π, an excellent fill factor of >95% and response times of up to 500 Hz. However, the current resolution of 240×240 pixels is too low to be useful in the digital holographic microscopy and there is no mass market behind this technology that would encourage a rapid development to improved devices at low costs [40].

2.6 Micromanipulation with Light

As early as 1969, the physicist Arthur Ashkin discovered that it is possible to accelerate and trap particles by light radiation pressure. In his experiments he used transparent latex spheres with diameters in the range of 1–4 wavelengths. He was able to show that these were pulled to the beam axis and accelerated in the beam direction during an exposure with a tightly focused laser beam [49].

Since the late 1990s, the possibilities provided by new digital holographic technologies were harnessed to create phase and/or modulated light fields at video rate. Therefore, holographic optical traps (HOTs) deliver powerful means to produce extraordinary, almost arbitrary beam configurations dynamically [50].

The principle of one single optical tweezer has been well known for more than 20 years. The intrinsic gradient of intensity of a TEM_{00} laser beam is utilized to exert an effective force to preferably dielectric particles with a size in the range of tens of nanometres to many micrometres [51]. As one optical trap or optical tweezer is comparatively simple to implement, but relatively powerful in its capabilities and applications, single optical tweezers quickly became significant tools in medical as well as biological sciences [52]. Nevertheless, a single tweezer is confined to the manipulation of one object and not to multiple simultaneously. Therefore, many approaches and concepts allowing the manipulation of two or more objects simultaneously have been investigated. The simplest concept is given by coupling two laser beams into the same optical beam path of a microscope [53]. In case the two beams are controlled separately before they are combined, both optical traps can be modified independently. The aforementioned approach is known as spatial multiplexing. An alternative concept is time multiplexing of two or more optical traps [54]. By this, typically one single laser beam is changed in its direction by a steering mirror like a piezo or galvo mirror or by an acousto-optical system that usually comprises vibrating crystals [55]. In applications, the laser beam is deflected to the desired position of an optical trap, held there for a desired duration and deflected to a following position in 3D space. The corresponding method is known as time multiplexing or time sharing. Each method - time and spatial multiplexing - have their advantages and inconveniences. The spatial multiplexing needs a separate single beam per trap and therefore the effort increases with each trap. Consequently, commercially available optical trap systems that are based on spatial multiplexing are limited to only a few, typically two independent traps (see also Sect. 3.1). Contrary, temporal multiplexing is realizable by partitioning of the trapping duration and laser power on the maintained optical trap positions on a very small temporal scale with high repetition rates. Therefore, trap stiffness is significantly reduced especially for higher number of traps.

The most flexible method to generate multiple optical traps in the focus plane of high quality optics such as a microscope objective is given by Holographic Optical Tweezers (HOTs) [56]. Therefore, a hologram in reflection or transmission geometry is placed in the optical path and thus read out by a reference wave. The numerical, digital hologram is pre-calculated with a computer and thus almost arbitrary intensity or phase patterns in the trapping plane are in principle possible, but primarily limited by diffraction and hardware restrictions. Typically, the hologram is placed in a Fourier plane of a microscope objective or a lens with respect to the trapping plane.

Multiple optical traps with the (digital) HOT-approach solely represent a specific case of possible complex trapping configurations or geometries. In order to exert high optical forces, significantly higher laser powers are required in the trapping plane since the optical gradient force also scales with intensity. Consequently, for a sufficient applicability of holographic optical tweezers, a high diffraction efficiency is mandatory and therefore typically phase holograms are utilised, preferably realised with phase only spatial light modulators. Alternatively, the needed hologram can be generated, for instance, by lithography techniques [56, 57].

Contrary, a significantly flexible approach with regard to space and time is given by the aforementioned dynamic holographic optical tweezers [58, 59]. Here, the hologram is generated by a computer controllable spatial light modulator (SLM) which enables the user to modify the trapping geometries by the display of a new hologram on the SLM that is located in a Fourier plane. Mechanically, the setup performs no movements. Nevertheless, the calculation of a hologram can be relatively demanding and time-consuming for a computer since also slightly different trapping patterns require the calculation of the complete new hologram. Consequently, the calculation of the holograms can be a limiting factor and her confinement for the dynamics in real-time applications. In comparison to direct imaging methods, a tweezer setup with a spatial light modulator can represent a limiting factor with

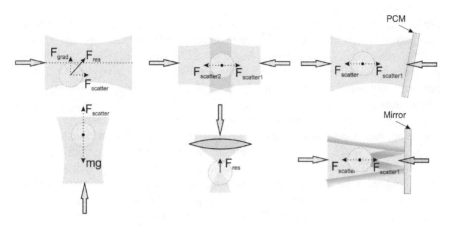

Fig. 2.13 Forces of a focused expanded beam, a two-beam trap, a trap with a phase-conjugating mirror (upper line) as well as a gravity trap, a focusing trap and a twin trap (lower line)

regard to the maximum trapping force and laser power as diffraction losses cannot be entirely reduced. Furthermore, the transmission and reflection coefficient put a limit on the maximum number of utilisable traps, the maximum trapping power and the damage threshold of the SLM concerning the maximum intensity that can be handled by the device.

In order to trap particles in space, several methods are available (see Fig. 2.13). For example, two opposed rays on one common optical axis can be used to hold a particle in space, since the two beams move the particle to the optical axis but the acceleration forces in beam direction add up to zero. This is also possible with a phase-conjugating mirror (PCM). A divergent beam is not reflected divergent here like on an ordinary mirror, but convergent and in the same direction from which it has come [60]. Gravity traps represent an alternative [61]. Here a particle is radiated from below and the power is adjusted to compensate the gravitational force on the particle, therefore trapping it at a fixed location. A third trap type focuses the laser beam with a high aperture lens (e.g. a microscope objective), which results in a force towards the focus point [62].

For a theoretical description of the trap force the size of the trapped particle is of critical importance. There are two generally accepted models: the geometric optical model for particles which are significantly larger than the used wavelength ($d > 5\lambda$) developed by Ashkin, and the wave-optical model for much smaller particles ($d < 0, 2\lambda$). There are several theories to describe the trap forces in the area between these scales, but no universally accepted solution. After a descriptive analysis of the momentums at large particles, a more detailed description of the two scales follows.

2.6.1   Observation of the Momentum

Optical trapping of micro-particles is possible because light beams have a momentum that can be transferred to matter. Therefore these particles should be at least partially

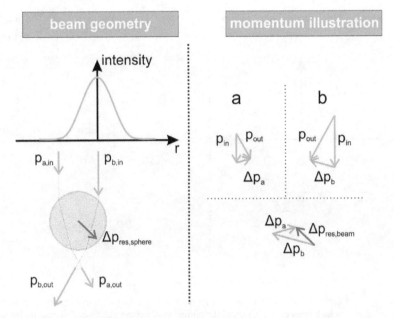

Fig. 2.14 Beam geometry and momentum-based analysis for a spherical transparent object outside the optical axis (decentered) of a collimated beam. Right: vectorial addition of momentum before and after transmission of the sphere for the rays a and b

transparent. If a beam impinges on the surface of e.g. a transparent sphere, it is partly reflected and partly transmitted. The transmitted beam is refracted, i.e. it changes its direction. The momentum is a vector quantity $\vec{p} = m \cdot \vec{v}$ and since the refracted light beam changes the direction, its momentum must change, too. According to the momentum conservation law, the sphere undergoes the same change in momentum in the opposite direction.

Two examples will illustrate the momentum changes. In an optical trap, a laser beam with an intensity profile which has an intensity gradient, for example a Gaussian intensity profile, is required. A bundle of infinitesimal beams from the beam centre has a larger impulse than a bundle from the edge region. If the partial beams a and b of a collimated laser beam hit e.g. a transparent sphere, they are partly refracted and thus change their direction (see Fig. 2.14). The momentum change of the beam $\Delta \vec{p}$ is derived from the vector subtraction of the momentum before entering the sphere $\vec{p}_{in}$ and the momentum after leaving the sphere $\vec{p}_{out}$. The resulting change in momentum of the entire beam $\Delta \vec{p}_{res,beam}$ follows from an addition of the momentum changes of all partial beams. Thus the entire beam experiences a change in momentum to the upper left. The sphere experiences the same change in momentum in the opposite direction $\Delta \vec{p}_{res,sphere} = -\Delta \vec{p}_{res,beam}$. Since the mass of the ball remains constant, the momentum change leads to a force, and thus an aligned velocity change of the sphere. Therefore, the sphere is moved to the highest intensity in the beam centre, but also in beam direction (see (2.73)).

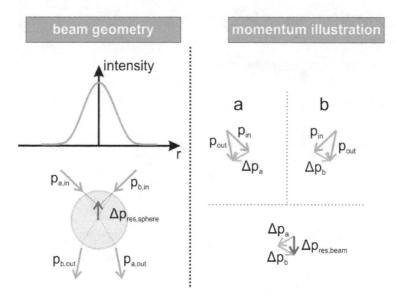

Fig. 2.15 For a focused laser beam results - analogous to Fig. 2.14 - a change in momentum of the spherical object, which is therefore driven to the optical axis and the location of maximum intensity (the focus). Right: vectorial addition of momentum before and after transmission of the sphere for the rays a and b

$$\frac{\mathrm{d}}{\mathrm{d}t}\vec{p} = \frac{\mathrm{d}}{\mathrm{d}t}(m \cdot \vec{v}) = \underbrace{\frac{\mathrm{d}}{\mathrm{d}t}m \cdot \vec{v}}_{0} + m \cdot \frac{\mathrm{d}}{\mathrm{d}t}\vec{v} = m \cdot \vec{a} = \vec{F} \qquad (2.73)$$

Analogously, this approach of the momentum is shown for a strongly focused laser beam in Fig. 2.15. For an easier representation the sphere is already on the optical axis of the beam, so the two partial beams have the same intensity and thus the same strength of momentum. The resulting change in momentum of the whole beam is pointing down, so the change in momentum of the sphere shows up to the laser focus. Thus the sphere moves to the laser focus and against the beam direction [63].

The scattering force F_{sca} accelerates the sphere in the beam direction and is produced by the absorption and reflection of light at its interfaces. It is proportional to the optical intensity, and points in the direction of the propagating light. The second force is the gradient force F_{grad} which results from the refraction of the light. It is proportional to the intensity gradient and points in its direction, provided that the refractive index of a particle is greater than that of the surrounding medium. If this is not the case, the force acts in the opposite direction and pushes the particles away from the location of highest intensity. Thus, a stable axial trap with a focused beam is only possible if the gradient force is stronger than the scattering force [63].

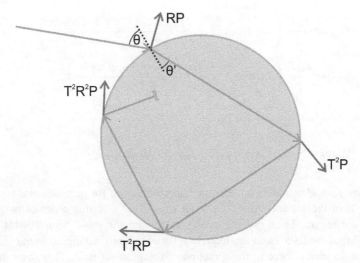

Fig. 2.16 Geometric illustration of refraction and reflection of a light beam on a sphere causing a resulting force on a approximately spherical particle

2.6.2 Geometric Optical Explanation - Mie Regime

This model is used to quantify the qualitative results of the previous chapter for particles that are larger than the wavelength of the utilized laser ($d > 5\lambda$). If a single beam with a power P hits a dielectric sphere at an incident angle θ, it is partly reflected and partly refracted according to Snell's law of refraction:

$$n_M \sin \theta = n_O \sin \theta', \tag{2.74}$$

where n_M, n_O denote the refractive index of a medium, respectively, of a trapped object. The refracted part of a beam is again partially reflected and partially refracted at the next interface between sphere and medium, etc. (see Fig. 2.16).

The force acting on an object is derived from the sum of the first reflected part with the power RP and all other refracted partial beams with decreasing power $T^2P, T^2RP, \ldots T^2R^nP$. Where R and T represent the Fresnel amplitudes of reflection and transmission coefficients.[4] The forces acting on the sphere are derived from the momentum difference between the incident and emerging light field. Here, the components in the beam direction are called the scattering force F_{sca} and the vertical components the gradient force F_{grad}. The total forces are calculated by integration of the transferred momenta of all partial beams [64]:

---

[4]The Fresnel formulas provide a quantitative description of the reflection and transmission of an electromagnetic wave at a plane interface. For a more detailed explanation it is referred to the corresponding literature [42, pp. 196ff.].

$$F_{\text{sca}} = \frac{n_M P}{c} \left(1 + R\cos 2\theta - \frac{T^2(\cos 2(\theta - \theta') + R\cos 2\theta)}{1 + R^2 + 2R\cos 2\theta'} \right)$$

$$F_{\text{grad}} = \frac{n_M P}{c} \left(R\sin 2\theta - \frac{T^2(\sin 2(\theta - \theta') + R\sin 2\theta)}{1 + R^2 + 2R\cos 2\theta'} \right).$$

(2.75)

2.6.3   Wave Optical Analysis - Rayleigh Regime

The wave-optical model is an adequate approximation for particles that are much smaller than the wavelength of the laser ($d < 0.2\,\lambda$). In this order of magnitude light is not regarded as a ray, but as an electromagnetic wave. Simplified it can be assumed that the laser radiation induces a variable electrical dipole in the particles, which undergoes a force in the direction of the gradient field. There is a gradient force F_{grad} in the direction of the intensity gradient ∇e^2 which is proportional to the gradient and the radius r_O of a particle with a particle polarizability α_P [63]:

$$F_{\text{grad}} = -\frac{n_M}{2} \alpha_P \nabla E^2 = -\frac{n_M^3 r_O^3}{2} \left(\frac{\tilde{n}^2 - 1}{\tilde{n}^2 + 2} \right) \nabla E^2.$$

(2.76)

The relative refractive index $\tilde{n}$ is obtained from the ratio of the refractive indices of the object and medium $\tilde{n} = n_O/n_M$. The equation shows that the gradient increases with the size of the particle. The scattering force F_{sca}, which accelerates a particle in beam direction, is given by:

$$F_{\text{sca}} = \frac{I_{\text{tot}}}{c_0} \frac{128\pi^5 r_O^6}{3\lambda_M^4} \left(\frac{\tilde{n}^2 - 1}{\tilde{n}^2 + 2} \right)^2 n_M.$$

(2.77)

Where I_{tot} represents the total intensity of a laser beam, c_0 the speed of light in vacuum and λ_M the material-dependent wavelength. In this model for a stable three-dimensional capture the gradient force also has to be greater than the scattering force [63].

2.6.4   Features and Influences of Optical Traps

Optically trapped particles are not rigidly anchored in space, but rather elastically retained to the place of maximum intensity. If all the trap forces are in equilibrium the particles rests at this location, but it can still rotate around its centre. If it is deflected out of the position of equilibrium of forces there is a restoring force that in first order increases approximately linearly with the deflection:

$$\vec{F}_{\text{res}} = -a_x \vec{x}.$$

(2.78)

The constant a_x is a rate for the lateral stiffness of the trap. In traps with a higher stiffness, a greater force is required to deflect a particle the same distance from the point of equilibrium of forces. Analogously, a_z represents the stiffness of the trap in the axial direction. The strength of force depends on the quality Q, the laser power P and the refractive index of the ambient (or biologic cells' nutrition) medium n_M [65, p. 260]:

$$\left|\vec{F}_{\text{res}}\right| = \left|\vec{F}_{\text{grad}} + \vec{F}_{\text{sca}}\right| = \frac{n_M P}{c_0} Q, \quad Q \in [0, 1]. \tag{2.79}$$

The dimensionless parameter for the quality of the trap is subject to various factors that will be explained in more detail in the following:

Numerical Aperture: This parameter describes the ability of an optical element to focus light and is obtained from the half angle of aperture α, and the refractive index of the immersion medium n_{imm} and is given as follows: $NA = n_{\text{imm}} \cdot \sin \alpha$. A microscope objective with a larger NA may therefore image a beam with a larger opening angle and thus has a higher resolution according to Abbe's theory: $d_{\text{res}} = \lambda/(2 \cdot NA)$. An increase in the NA results in a reduction of the spot size as well as an increase in the gradient force because the intensity gradient in the focal region is higher. Since it is important for a stable axial trap that the gradient force is greater than the scattering power, a high NA is essential. This is achieved usually with a NA ≥ 1. Especially the marginal rays of the entrance pupil of the objective contribute to the axial gradient force and therefore should have a high intensity. In a typical Gaussian intensity profile this is not the case. For more homogeneous illumination it is therefore required to widen this beam further than the entrance pupil.

Capture geometry: In general simple focal points are used for traps that have a Gaussian intensity distribution in the lateral and axial direction. However, for special applications it may be necessary to use special intensity geometries for trapping. For example with toroidal ("doughnut") traps particles such as air bubbles can be trapped, which have a lower refractive index than the surrounding medium and would therefore be repelled by a simple trap [66] (see also Sect. 4.8).

Aberrations: Optical aberrations have to be minimised as far as possible, since they usually reduce the intensity gradient and thus the resultant trap force. The aim is a diffraction-limited performance of the system, i.e. the imaging quality is not limited by the aberrations of the system, but only by the effect of diffraction.

Direction of polarisation: The lateral trap force is dependent on the direction of polarisation of light in the trap. It is greater parallel to the direction of polarisation than perpendicular thereto. The direction of polarisation has no direct effect on the axial trap force [67].

Refractive index ratio: The ratio of the refractive index of the particle to the refractive index of the ambient medium ($n = n_O/n_M$) has an influence on the maximum achievable trap force, what could be experimentally demonstrated [64]. The best trap forces can be achieved in the range $n = 1.2 - 1.5$ where most biological samples have their nutrition medium. With an increasing refractive index ratio the achievable trap forces decrease again because the scattering power increases faster relative to the gradient force (see (2.75)). The refractive index ratio determines the deflection angle θ of a light beam at the interface between the medium and particles.

Radius of the object: The influence of the radius depends on the underlying computational model. For large particles in the geometric-optical range, the specific trap force is independent of the object size. By contrast, in a wave-optical model the gradient force increases exponentially by the cube of the object radius (see also (2.76)). Therefore, particles of diametre $d < 0.2\,\lambda$ show a strong dependence of the object size which fades to independence in the geometric-optical range ($d > 5\,\lambda$) [68].

Object form: The highest trap forces are obtained for objects that have an ideal spherical shape. The more the actual shape differs from this, the greater is the scattering on the surface. This reduces the gradient force, resulting in a reduction of the trap force, since the ratio between the gradient and scattering force becomes smaller.

Therefore, according to (2.79), an increase of the trap force can be primarily achieved through an improvement of the quality of the trap and an increase of the laser power. However, it should be noted that the laser power cannot be increased arbitrarily since damage to the biological samples is undesirable. Damage can be caused by action of heat as a result of absorption and by photochemical processes in the cell. In order to minimize the damage to the cell caused by absorption, the used laser wavelength can be modified. The transparency of biological cells is strongly dependent on the wavelength of the light. Biologically colouring material such as hemoglobin or other cytochromes absorb near-infrared light much less. The absorption in water, however, increases sharply with increasing wavelength and has its maximum at about $3\,\mu m$. A good compromise is a wavelength in the region of around $1064\,nm$. There is a local minimum of the absorption coefficient of water, and the absorption of biological cell components is significantly reduced compared to the visible region [65, pp. 257f.].

2.6.5 Algorithms for Optical Trap Patterns in the Fourier Plane

An often utilized algorithm for calculating the phase hologram is a non-iterative "gratings and lenses" algorithm (OpenGL code listed in Appendix A.1.3). The microscope lens performs a Fourier transform of the hologram. Thus, in order to obtain the hologram, an inverse Fourier transform of the desired intensity pattern on the object plane has to be calculated. The utilised algorithm makes use of numerical tracing optical diffraction gratings and lenses to determine the inverse Fourier transform of the desired trap pattern. A trap spot can be displaced with a diffraction grating in the lateral direction. Thereby the angle of the grating determines the direction of deflection and the frequency of the grating the amount of deflection. An axial displacement is achieved by a lens term. The required phase distribution in the hologram for a lateral displacement of the trap spots around $(\Delta x, \Delta y)$ is given by:

$$\Phi_{grating}(x_h, y_h) = \frac{2\pi}{\lambda f}(x_h \Delta x + y_h \Delta y). \qquad (2.80)$$

Where x_h, y_h are the coordinates in the hologram plane, λ the wavelength of the laser beam, and f the effective focal length of the lens [69]. The equation for calculating the lens term for an axial displacement Δz of the trap is:

$$\Phi_{lens}(x_h, y_h) = \frac{2\pi \Delta z}{\lambda f^2}(x_h^2 + y_h^2). \qquad (2.81)$$

Furthermore, an optical vortex with a helical phase pattern can be generated using Laguerre–Gaussian beams, which have an orbital angular momentum and which thus can exert optical torque (see also Sect. 4.8.1, [70, p. 111]). The phase pattern required for generating such a beam is given by:

$$\Phi_{vortex}(x_h, y_h) = -l \tan^{-1}(y_h/xh) + \pi\theta[-L_p^{|l|}(2(x_h^2 + y_h^2)/w_0^2)]. \qquad (2.82)$$

Here the parameters p and l represent the radial and azimuthal mode, $L_p^{|l|}(x)$ the generalised Laguerre polynomial, θ a uniform step function and w_0 the beam waist. Only the parameter for azimuthal mode ($l = L$) is left variable in the algorithm implemented for the experimental investigations in Sect. 4.8.3. The radial mode (p) remains unchanged at 0.

If a trap is simultaneously deflected laterally as well as axially and, in addition, has an optical vortex, the corresponding hologram results from the sum of the (2.80) to (2.82). In order to display it on the SLM the hologram has to be folded in the interval from 0 to 2π:

$$\begin{aligned} \Phi_h(x_h, y_h) &= \Phi_{grating} + \Phi_{lens} + \Phi_{vortex} \\ &= (\Phi_{grating} + \Phi_{lens} + \Phi_{vortex}) mod\, 2\pi \end{aligned} \qquad (2.83)$$

The hologram for N traps results from the superposition of the individual holograms from (2.83):

$$\Phi_{ges} = arg\left(\sum_{n=1}^{N} \exp(i\Phi_{h,n})\right) = \tan^{-1}\left(\frac{\sum_{n=1}^{N} \sin(\Phi_{h,n})}{\sum_{n=1}^{N} \cos(\Phi_{h,n})}\right). \qquad (2.84)$$

By means of the intensity weighting parameter I_w of an implemented software for the experimental investigations, the weighting of individual traps can be influenced in the superposition, whereby different light intensities of traps are realised. Figure 2.17 illustrates examples of some hologram patterns.

Fig. 2.17 Phase holograms modulo 2π: A trap laterally deflected in x and y (top left), two traps laterally deflected in x and y (top centre), an axial deflected trap (top right), undeflected Laguerre-beam ($L = 8$) (bottom left) and two laterally and axially deflected traps (bottom right)

2.6.6   Calibration of the Trap Forces

Many applications of optical tweezers require quantitative information about the trap forces. Since the existing models to calculate the trap force are not sufficiently accurate or contain too many unknown variables, an empirical characterisation and calibration of the trap forces is required. Thereby, two different principles are used which make use of different physical effects for determination of the trap force: the fluid friction and Brownian motion.

If a particle moves in a viscous medium, the force of fluid friction acts against the movement direction. The strength of this force can be calculated using the Stoke's law, which describes the friction force experienced by a sphere of radius r which moves with the velocity v in a homogeneous medium of dynamic viscosity η:

$$F_R = 6\pi r\eta v. \tag{2.85}$$

At a constant velocity of a particle, induced by an optical trap, the trap force corresponds to the force acting against the force of fluid friction. With stepwise increase in the speed of movement, the maximum trap force is exceeded if the trapped particle escapes from the trap. Instead of moving the optical trap, the medium can also be moved, for example, by displacing the object table or a defined volume flow in a corresponding fluid chamber. The calculation of the particle velocity is possible via image processing [65, pp. 267ff.].

The evaluation of the trap force by Brownian motion takes advantage of this stochastic motion, which microscopic particles carry out. When a particle is captured, its movement is reduced proportional to the trap force. By detecting this reduced position variation, the magnitude of the trap force can be concluded through a cut-off frequency procedure or Boltzmann statistics. To perform a statistical analysis of the object positions, the positional variations of the particle have to be measured over a certain period of time which increases with growing size of the object [48, pp. 72ff.]. In this work the trap force is determined by the escape method since it features a high reliability at a short measurement time.

2.7 Dynamic Holography for Optical Micromanipulation

For many applications of optical tweezers a dynamic generation of an arbitrary amount of spots that can be moved independently in three-dimensional space is of great advantage. This can be realised with a structure that uses the principle of holography, to diffract the light to the desired intensity pattern. The central element of the system is an SLM, which can change the phase of a laser beam from 0 to 2π in each pixel separately. Therefore the SLM is read out with a collimated laser beam, and is coupled into a microscope (see Fig. 2.18). The beam diameter is adjusted by a telescope made of a coupling and a tube lens at the entrance pupil of the microscope objective. A dichroic mirror is used to separate the laser beam of the illuminating beam path, so that the enlarged image of the object plane can be digitised using a camera. From a desired intensity distribution a computer-generated hologram is calculated on the computer and displayed on the SLM. In order to realise real-time video calculation, this is outsourced to the graphics card of the computer. The SLM is located in the Fourier plane of the object plane meaning the microscope objective produces an optical Fourier transformation of the modulated wave front. This finally leads to the desired reconstruction of the hologram in focal respectively the object plane.

The positioning of the trap can be done in three dimensions. For a single trap which is deflected from the focus point a sawtooth-shaped diffraction grating can be used as a hologram. The lateral position can be adjusted by the spatial frequency and the angle of the grating. For a displacement of the focus in the axial direction the hologram can be superimposed by a phase pattern in accordance with a Fresnel zone lens. The hologram for several traps results from the superposition of the individual hologram traps [71].

In summary, holographic systems have the following advantages for optical micro-manipulation:

- compact and simple setup, since the trap will not be positioned mechanically and complex and error-prone controls disappear

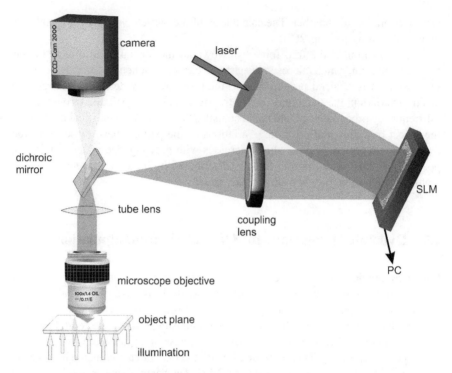

Fig. 2.18 A laser beam is reflected at an SLM which is computer controlled and able to modulate the phase of the laser beam. The modulated beam is coupled into a microscope and Fourier transformed by a microscope objective. A dichroic mirror separates illumination and trap beam whereby the illumination beam can be imaged with a camera

- variable number of traps possible, limited only by the maximum laser output power and the local bandwidth product of the SLM
- moving of the particles in three-dimensional space
- dynamic change of the trap positions in real time
- compensation of aberrations with the SLM
- high and repeatable accuracy of positioning, since this is entirely digitally controlled

A disadvantage of the principle is a higher power loss and therefore a need for a stronger trap laser in comparison to systems of conventional optical tweezers based on one single trap. These power losses are caused by the non-perfect reflectivity of the SLM, additional optical elements such as polarisation filters and especially by expanding the laser beam to homogenize the intensity profile. More power losses are induced by the use of non-ideal SLMs, i.e. a no linear phase shift or a maximum phase shift less than 2π, as this causes zero diffraction orders or ghost traps.

2.8 Applications of Optical Tweezers

Optical tweezers have numerous applications in medical cell research, biophysics, chemistry and micro system technology [50]. Since a list of possible applications is beyond the scope of this work, some examples of applications are presented in the following.

Especially in case of handling living cells, the method represents a major benefit due to its minimally invasive nature. In biophysics, optical tweezers are used to measure mechanical properties of cells and biological structures. By exerting directed forces in the range of pN, it is for example possible to measure the elasticity of DNA [72]. The DNA is therefore fixed at one end and at the other end attached to a polystyrene sphere. By selected trapping and deflection of the sphere, the forces can be determined which the DNA can provide. In a similar manner, the forces of biological compounds - such as antibody/antigen compounds - can be measured [73].

Another application in cell research is the cell sorting and determined positioning. In that way it is possible to transport cells in microfluidic systems into different chambers where they may be exposed to certain environmental conditions. For example, the effects of medicine on biologic cells can be studied better or the cells may be exposed to certain specific fluorescent markers [74].

In combination with another laser for selectively destroying biological material (laser scalpel), optical tweezers are also used for micro-cloning of DNA. Therefore, a selected chromosome is isolated with the optical tweezers and then the component, which contains the required information (for example, a particular gene) is separated with a laser scalpel. The target fragment is then transported by means of optical tweezers in a region where it is reproduced. In this way, only micro-clones of the desired DNA segment emerge [75].

One application specifically for holographic optical tweezers, wherein a plurality of particles have to be moved independently, is the drive of micro-optomechanical pumps. Therefore special holograms generate multiple optical vortex in the object plane, making it possible to specifically transfer an angular momentum to trapped particles. Toroidal intensity patterns are formed, which catch particles in the "intensity Ring" and additionally rotate them within. With an array of such special patterns a volume flow can be established and directed (see Fig. 2.19 and Sect. 4.8) [76].

2.9 Diffractive and Non-diffractive Beam Types

Over a long period of time it was accepted by the scientists that a wave of finite extent, which propagates in space, necessarily follows Huygen's principle. In other words, it is impossible to produce a wave which does not change in form over a longer distance. According to Huygens, each wave will be diffracted as it passes through a barrier or enters another medium.

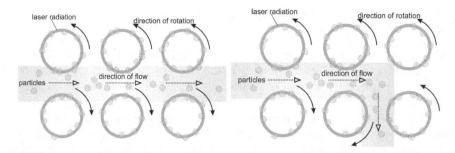

Fig. 2.19 By rotation of particles a micro-pump can be generated, which allows small volume flows

In this regard, each point of the refracted wave can be regarded as an infinitesimal small source of a new wave (elementary wave). The shape of each new wave is spherical, as long as the size of the barrier (often an aperture) is of significant extent compared to the wavelength. The spherical part represents the diffraction of the wave, because only a completely flat wavefront is considered as non-diffractive. The new waves interfere with each other and thus generate an interference pattern.

It should be noted that it is not feasible to generate a homogeneous plane wave with a point source (e.g. a laser), since the laser acts as an aperture and the Gaussian beam intensity distribution will always remain, which will later be shown in theoretical relations. Therefore, the beams sometimes are not simply called Bessel or Laguerre-beams but Bessel–Gaussian and Laguerre–Gaussian beams. Nevertheless, the diffractivity can be significantly reduced in specific areas with the help of a SLM by modulation of the incident wavefront.

2.9.1   Gaussian Beams

Taking the just made considerations into account, the Gaussian beam (also known as TEM_{00} mode) is now assumed, which is a widely used description of light propagation, using the methods of radiation and wave optics. Most lasers produce a wave that can be described by a Gaussian beam since its intensity is subject to the bell-shaped curve (Fig. 2.20).

The Gaussian beam is described by the following equation in cylindrical coordinates [70]:

$$E(r, z) = E_0 \frac{\omega_0}{\omega(z)} e^{-\left(\frac{r}{\omega(z)}\right)^2} e^{-ik\left(\frac{r^2}{2R(z)}\right)} e^{i(\zeta(z) - kz)} \qquad (2.86)$$

with the amplitude E_0, the wave number $k = \frac{2\pi}{\lambda}$, the beam radius $\omega(z) = \omega_0 \sqrt{1 + (\frac{z}{z_0})^2}$, the radius of curvature $R(z) = z(1 + (\frac{z_0}{z})^2)$ and the Gouy phase $\zeta(z) = \arctan(\frac{z}{z_0})$. The Rayleigh length $z_0 = \frac{\pi \omega_0^2}{\lambda}$ represents the distance in which

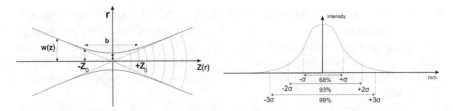

Fig. 2.20 Gaussian beam with its specific parameters (left) and the bell-shaped Gaussian curve with indicated intersections for the standard deviation (right)

the beam diameter increases its waist (the narrowest cross-sectional area) by a factor of $\sqrt{2}$. Hereby, it can be clearly seen that the beam radius strongly depends on its z position and thus has to be referred to as diffractive. (For more information see Appendix B.3.10.)

This equation, as well as any other equation describing a wave front, is derived from the Helmholtz equation (see also Appendix B.3), also known as the wave equation:

$$\nabla E(\vec{x}) = -k^2 E(\vec{x}) \,, \tag{2.87}$$

with the wave number $k = \frac{\omega}{c}$. The Helmholtz equation is derived from the Maxwell's equations (see also Appendix B.3), developed by James Clerk Maxwell in the 1860s, for an electromagnetic wave in isotropic, linear, time-independent and homogeneous materials. When making use of a laser, the following approach is utilised for the electric field [77]:

$$E(\vec{x}) = E_0 X(x, z) Y(y, z) e^{ikz} \tag{2.88}$$

All other beam types can be determined by the solutions of the wave equation. In case of a Gaussian beam, the following equations are a possible solution:

$$
\begin{aligned}
X_m(x, z) &= \sqrt{\frac{\omega_0}{\omega(z)}} H_m \left(\frac{\sqrt{2}x}{\omega(z)} \right) \exp \left(-\frac{x^2}{\omega^2(z)} - i\frac{kx^2}{2R(z)} + i\frac{2m+1}{2}\zeta(z) \right) \\
Y_n(x, z) &= \sqrt{\frac{\omega_0}{\omega(z)}} H_n \left(\frac{\sqrt{2}y}{\omega(z)} \right) \exp \left(-\frac{y^2}{\omega^2(z)} - i\frac{kx^2}{2R(z)} + i\frac{2n+1}{2}\zeta(z) \right).
\end{aligned}
\tag{2.89}
$$

H_i denotes the Hermite polynomials, where $m = n = 0$ applies to a Gaussian beam.

2.9.2 Bessel Beams

In 1987 [78, 79] has presented a solution to the wave equation (2.87), which no longer depends on the Rayleigh length $z_r = \pi \omega_0^2 / \lambda$ and has a sharp intensity distribution. Therefore these beams are denoted as non-diffracting and in this case called

Bessel beams. An outstanding property of the Bessel beams is also the opportunity to transmit a torque to the object, or to create a channel for cell transport by an arrangement of multiple beams (see also Fig. 2.19). Such rays are referred to as OAM-Bessel beams (Orbital Angular Momentum).

The scalar, non-diffractive solution of the Helmholtz equation in Cartesian coordinates with $z \geq 0$ is given by [78]:

$$E(x, y, z \geq 0, t) = \exp\left[i(\beta z - \omega t)\right] \int_0^{2\pi} A(\phi) \exp\left[i\alpha(x \cos(\phi) + y \sin(\phi))\right] d\phi$$

(2.90)

with an arbitrary complex ϕ-dependent amplitude $A(\phi)$ and the wave number $\beta^2 + \alpha^2 = (\omega/c)^2 = k$. The first exponential term describes the temporal and local propagation in z direction with the given frequency ω and the wave number β. The second exponential term represents the changing cross-section surface in the transverse plane. The intensity profile in propagation direction remains unchanged [78]:

$$I(x, y, z \geq 0) = 1/2|E(r, t)|^2 = I(x, y, z = 0),$$

(2.91)

if β is real, i.e. the normals of the wave front correspond to the z-axis.

To obtain a non-diffractive *Bessel*-geometry, the ϕ-dependency is eliminated, whereby a constant wave front is constructed. Carrying out a simple transformation and a normalisation of the integral term to 2π, this results - with the help of (2.90) - in the following expression for the electric field:

$$E(r, t) = \exp\left[i(\beta z - \omega t)\right] \int_0^{2\pi} \exp\left[i\alpha(i \cos(\phi) + y \sin(\phi))\right] \frac{d\phi}{2\pi}$$

$$= \exp\left[i(\beta z - \omega t)\right] J_0(\alpha \rho)$$

(2.92)

Here $\rho = x^2 + y^2$ is the radial distance from the beam centre and J_0 a Bessel function of zero order. If the argument of the Bessel function is zero, a simple plane wave front is formed as the function itself is 1 (Fig. 2.21).

If $0 < \alpha \leq \omega/c$, a non-diffractive beam with a sharp maximum of intensity is formed which decreases proportionally to $\alpha \rho$ and has only the width $3\lambda/4$.

The *Bessel function* is generally defined as the solution of Bessel's differential equation [80, 81]:

$$x^2 \frac{d^2 f}{dx^2} + x \frac{df}{dx} + (x^2 - n^2)f = 0,$$

(2.93)

where n is a real or complex number, which determines the order of the Bessel function. The Bessel function can be written in several ways - e.g. as an integral or series representation [80, 81]:

Fig. 2.21 Bessel functions of the first 5 orders

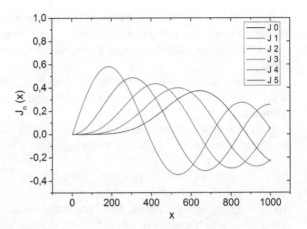

$$J_n(x) = \frac{1}{\pi} \int_0^{\pi} \cos(n\tau - x\sin(\tau))d\tau \qquad (2.94)$$

$$J_n(x) = \frac{1}{2\pi} \int_{-\pi}^{\pi} e^{-i(n\tau - x\sin\tau)}d\tau \qquad (2.95)$$

or

$$J_o = \sum_{k=0}^{\infty} (-k)^2 \frac{(1/4z^2)^k}{(k!)^2} \qquad (2.96)$$

To produce a Bessel beam or rather a Bessel–Gauss beam [82] since the natural intensity distribution of an applied laser beam is usually Gaussian, either an axicon (see Fig. 2.22, [83]) or a spatial light modulator can be used.

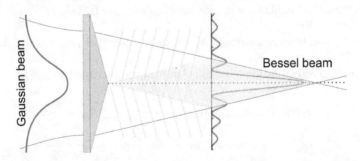

Fig. 2.22 The creation of a Bessel(-Gauss) beam with an incident Gaussian beam and a following axicon (blue glass cone) generating a corresponding profile of the intensity distribution on a screen (right)

The production of the beam by means of SLMs can be accomplished in two ways: Through an axicon phase pattern, which implies a simple Bessel beam, or through an annular aperture, which introduces, among other things, an angular torque to the wave.

The generation of an axicon on the SLM is similar to the generation of "Lenses" in the "Gratings and Lenses"-algorithm (2.81) and is only mentioned briefly in the following.

If the magnitude of the complex amplitude term $|A(\phi)|$ from (2.90) is set to 1 and the complex amplitude function of the hologram is given by

$$t(\rho, \phi) = A(\phi)e^{i(2\pi\rho/\rho_0)} \ , \tag{2.97}$$

whereas $t(\rho, \phi) = 0$ for $\rho > R$, and $A(\phi)$ is the same amplitude as in (2.90), with $\alpha = 2\pi/\rho_0$, which leads to $A(\phi) = e^{i(2\pi\rho/\rho_0)}$, then the wave equation is according to Durnin a Bessel function of the first kind with the following function for the phase hologram [84]:

$$\Psi(\rho, \phi) = n\phi + 2\pi\rho/\rho_0 \tag{2.98}$$

2.9.3   Superposition of Bessel Beams

For a generation of Bessel beams with a SLM a ring hologram is transmitted to the modulator and which is illuminated by a Gaussian beam of a laser. Thus, in practice an annular aperture is simulated with the aid of the modulator. In the following the realisation of an OAM Bessel beam is illustrated. All assumptions and procedures of this method can be reduced to the simple Bessel beam, without an angular momentum, by using only one annular aperture. As a result of the incident Gaussian beam, three zones arise in certain distances similar to the realisation using an axicon:

I. the near field **NF** which has usually highest relevance as this is the previously described Fourier or focal plane [85],
II. the intermediate field as a transition between near and far field **IF** and
III. the far field **FF**.

The generated hologram on the SLM with two annular apertures is as well depicted in Fig. 2.23.

The transmission function of the ring slits can be described by (2.99).

$$\begin{aligned} t_1(r, \phi) &= \exp(il\phi) \text{ for } R_{OC} - \frac{\Delta}{2} \leq r \leq R_{OC} + \frac{\Delta}{2} \\ t_2(r, \phi) &= \exp(-il\phi) \text{ for } R_{IC} - \frac{\Delta}{2} \leq r \leq R_{IC} + \frac{\Delta}{2} \end{aligned} \tag{2.99}$$

and is supposed to be $t_{1,2}(r, \phi) = 0$ elsewhere, with the average radius of the respective annular aperture R describing the inner (IC) and outer circle (OC), the width or

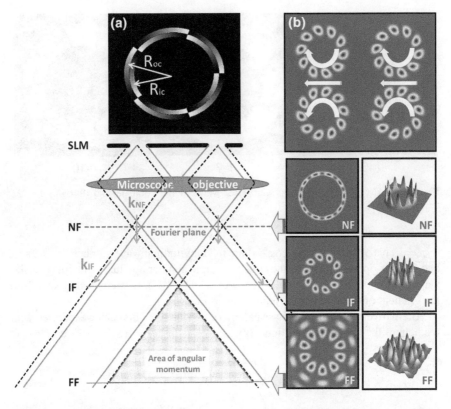

Fig. 2.23 Double slit annular aperture phase hologram modulo 2π displayed on an SLM in **a** with the characteristic parameters for generating a Bessel beam with an angular momentum. Below: Sketch of different Zones of a Bessel beam with the near (NF), intermediate (IF) and far field (FF). Right (**b**): Corresponding theoretical normed intensity distributions in the specified different planes

thickness of the rings Δ, the azimuth angle ϕ and the angular frequency l expressed by the amount of 2π transitions.

In order to determine the wave function (Figs. 2.24, 2.25, 2.26 and 2.27) in the three zones, a number of simplifications can be made. Since the width of the annular aperture is in the order of some µm, the incident Gaussian beam is assumed to be a plane wave $\exp(i(k_z z - \omega t))$ with the longitudinal wave number k_z and the propagation direction z. The time dependence of the exponential term ωt is also negligible because the plane wave remains uniform in the xy-plane over the entire time.

Near field According to the precedent simplifications, the following field distribution is visible in the near-field NF ($z \approx z_0$) [85]:

$$A^{\mathrm{NF}}(r, \phi, z_0) = \exp(il\phi) \exp(ik_{1z}z_0) \text{ for } R_{OC} - \frac{\Delta}{2} \leq r \leq R_{OC} + \frac{\Delta}{2} \quad (2.100)$$

$$A^{\mathrm{NF}}(r, \phi, z_0) = \exp(-il\phi) \exp(ik_{2z}z_0) \text{ for } R_{IC} - \frac{\Delta}{2} \leq r \leq R_{IC} + \frac{\Delta}{2} \quad (2.101)$$

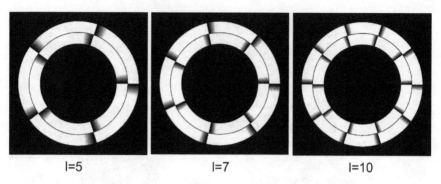

l=5   l=7   l=10

Fig. 2.24 OAM Bessel phase masks with respectively 5 (left), 7 (middle) und 10 (right) $0 \rightarrow 2\pi$ transitions per ring

$k_{1z} = k\cos(\alpha_1)$ and $k_{2z} = k\cos(\alpha_2)$ are the longitudinal wave numbers, $k = 2\pi/\lambda$ and α is the opening angle of the cone. A simple annular aperture, without intensity variation can be realised with $A^{\text{Ring}}(r) = \exp((\frac{-(r-R_0)}{\Delta/2})^n)$, whereby n corresponds to the gradient of the angle of the annular aperture.

Alternatively, the ring slit phase pattern of two annular slits can be expressed with the help of the top hat distribution $\pi(r)$:

$$E_{\text{NF}}(r, \phi, z_0) = \pi\left(\frac{r - R_{IC}}{\Delta}\right) \exp(il\phi) \exp(ik_{1z}z_0)$$

$$+ \pi\left(\frac{r - R_{OG}}{\Delta}\right) \exp(-il\phi) \exp(ik_{2z}z_0) \qquad (2.102)$$

Since the widths of the rings are small, the azimuthal phase rotation is in opposite direction and the optical system is linear, the term A^{NF} for two annular apertures (ring slits) yields

$$A^{\text{NF}}(r, \phi, z_0) \approx \exp\left(\left(\frac{-(r - R_0)}{\Delta/2}\right)^n\right) \left(\exp(il\phi)\exp(ik_{1z}z_0) + \exp(-il\phi)\exp(ik_{2z}z_0)\right)$$

$$(2.103)$$

by adding the field distributions of the individual aperture contributions and the use of a single average slit approximation ($R_{IC} \approx R_0 \approx R_{OC}$). To obtain the intensity distributions, the complex conjugate product $I = AA^*$ is calculated:

$$I^{\text{NF}}(r, \phi, z_0) = 4\cos^2\left(\frac{k_{1z}z_0 - k_{2z}z_0 + 2l\phi}{2}\right) \left(\cosh\left(\frac{2(r - R_0)}{\Delta}\right) - \sinh\left(\frac{2(r - R_0)}{\Delta}\right)\right)^{2n}$$

$$(2.104)$$

Hereby, the double phase modulation and thus the doubling of the number of the minima and maxima of the intensity along the ring profile becomes apparent. Written in a more compact way, the precedent relation may also be described analogue to (2.102) with a distribution:

$$I^{NF} = EE^* = 2\pi \left(\frac{r - R_{IC}}{d} \right) + 2\pi \left(\frac{r - R_{OC}}{d} \right)$$
$$+ \pi \left(\frac{r - R_{OC}}{d} \right) \pi \left(\frac{r - R_{IC}}{d} \right) 2 \cos(2l\phi + k_{1z}z_0 - k_{2z}z_0) \quad (2.105)$$

The torque on the particles in the near field is determined by the dependency of the angle of the z position (in this case, since $z = z_0$):

$$\frac{d\phi}{dz_0} = \frac{k_{2z} - k_{1z}}{2l} \quad (2.106)$$

Based on the Fig. 2.23 it can be seen that the two wave numbers k_{1z} and k_{2z} show in the same direction at the position $z = z_0$ and have the derivative 0.

Intermediate field In the transition area *IF* (intermediate field) the Fresnel approximation known from optics is used to calculate the field distribution.

$$A^{IF}(r, \phi, z) =$$

$$\frac{e^{ikz}}{i\lambda z} \int\limits_{0}^{2\pi} \int\limits_{R_1 - \Delta/2}^{R_2 + \Delta/2} A^{NF}(r, \phi, z) \times \exp \left(i\frac{k}{2z}(r_1^2 + r^2 + 2rr_1 \cos(\phi_1 - \phi)) \right) r_1 dr_1 d\phi_1$$

$$(2.107)$$

Here, the previously calculated near field is directly integrated at the slit over the slit surface and propagated with the help of the first exponential term to the correct position on the z-axis. This integral provides a Bessel–Gauss beam. The resulting equation has a strong resemblance to the solution of the Helmholtz equation in Sect. 2.9.1, with the exception of the Bessel function instead of the Hermite polynomials [85].

$$A^{IF}(r, \phi, z) = A^{BG1}(r, \phi, z) + A^{BG2}(r, \phi, z) = \left(\frac{1}{\sqrt{1 + (\frac{z}{z_r})^2}} \right) \times$$

$$\begin{pmatrix} \exp \left[i\left(kz - \frac{k_{1r}^2 z}{2k} - \Phi(z) \right) - \left(\frac{1}{\omega^2(z)} - \frac{ik}{2R(z)} \right) \left(r^2 + \left(\frac{k_{1r}z}{k} \right)^2 \right) + ik_{1z}z + il\phi \right] J_l \left(\frac{k_{1r}r}{1 + i(z/z_r)} \right) + \\ \exp \left[i\left(kz - \frac{k_{2r}^2 z}{2k} - \Phi(z) \right) - \left(\frac{1}{\omega^2(z)} - \frac{ik}{2R(z)} \right) \left(r^2 + \left(\frac{k_{2r}z}{k} \right)^2 \right) + ik_{2z}z + il\phi \right] J_{-l} \left(\frac{k_{2r}r}{1 + i(z/z_r)} \right) \end{pmatrix}$$

$$(2.108)$$

The parameters $\omega(z)$, $R(z)$ and $\Phi(z)$ are defined equivalently to the (2.86). The parameter J_l denotes the Bessel function of lth order and determine the transversal field distribution of the beam. The subscripts 1 and 2 denote the beam originating from the respective outer ring 1 and inner ring 2.

The calculation of the intensity is relatively complex at this point and can also not be found explicitly in literature. Complex calculations to simplify the expression without approximations lead to the following result:

$$
I_{\mathrm{IF}} = \sum_{i=1}^{8} \frac{1}{1+\left(\frac{z}{z_r}\right)^2} \left(\left(
\begin{array}{l}
\exp\left[-2\left(\frac{1}{\omega^2(z)}\right)\left(r^2 + \left(\frac{k_{1r}z}{k}\right)^2\right)\right] \left| J_l\left(\frac{k_{1r}r}{1+i\left(\frac{z}{z_r}\right)}\right) \right|^2 \\
+ \exp\left[-2\left(\frac{1}{\omega^2(z)}\right)\left(r^2 + \left(\frac{k_{1r}z}{k}\right)^2\right)\right] \left| J_{-l}\left(\frac{k_{2r}r}{1+i\left(\frac{z}{z_r}\right)}\right) \right|^2 \\
+ \exp\left[-2\left(\frac{1}{\omega^2(z)}\right)\left(r^2 + \left(\frac{k_{2r}z}{k}\right)^2\right)\right] \\
\cdot \left(J_l\left(\frac{k_{1r}r}{1+i\left(\frac{z}{z_r}\right)}\right) \cdot J_{-l}\left(\frac{k_{2r}r}{1+i\left(\frac{z}{z_r}\right)}\right)^* + J_l\left(\frac{k_{1r}r}{1+i\left(\frac{z}{z_r}\right)}\right)^* \cdot J_{-l}\left(\frac{k_{2r}r}{1+i\left(\frac{z}{z_r}\right)}\right) \right) \\
\cdot \cos\left(-\frac{k_{1r}^2 z}{2k} + \frac{k_{2r}^2 r}{2k} + k_{1z}z - k_{2z}z + l\phi\right)
\end{array}
\right) \right)
$$

$$(2.109)$$

Here, the angular rotation can be approximated according to (2.106), which is then not zero since the wave vectors cannot be considered parallel any more (see also Fig. 2.23).

Far field In the far field FF the situation is different and fields which experience the wave vektor of both rings are formed. First, the field distribution is calculated with the Kirchhoff diffraction integral:

$$
A^{\mathrm{FF}}(r, \phi, z) = \frac{-i}{\lambda z} \int_0^{2\pi} \int_{R_1-\Delta/2}^{R_2+\Delta/2} t(r, \phi_1)\, e^{\left[i\frac{k}{2f}\left(1-\frac{z}{f}\right)r_1^2\right]} \times e^{\left[-i\frac{k r r_1}{f}\cos(\phi_1-\phi)\right]} r_1 \mathrm{d}r_1 \mathrm{d}\phi_1
$$

$$(2.110)$$

The equation includes the transmission function of annular slits $t(r, \phi_1)$ and the exponential terms depend on structure-dependent parameters such as the focal length f, the distance z of the simulated rings of the regarded plane. In addition, the wave vectors, the radii of the rings and the angle dependence of the field distribution $\cos(\phi_1 - \phi)$ are considered in the exponential terms.

The superposition of the two waves of the two rings in the far field yield after the calculation of the integral:

$$
A^{\mathrm{FF}}(r, \phi, z) = J_l(k_{1r}r)\exp(il\phi)\exp(ik_{1z}z) + J_{-l}(k_{2r}r)\exp(-il\phi)\exp(ik_{2z}z)
$$

$$(2.111)$$

To obtain the intensity of the wave, again the complex conjugate product is calculated.

$$
I^{\mathrm{FF}}(r, \phi, z) = \propto J_l^2(k_{1r}r) + J_{-l}^2(k_{2r}r) + 2J_l(k_{1r}r)J_{-l}(k_{2r}r)\cos(k_{1z}z - k_{2z}z + 2l\phi)
$$

$$(2.112)$$

Because of the already used simplification of thin rings and approximately equal, transverse wave vectors $(k_{1r} \approx k_{2r} = k_r)$, the Bessel functions are simplified to $(J_l(k_{1r}r) \approx J_{-l}(k_{2r}r) = J_l(k_r r) \Rightarrow J_{-l}(k_{2r}r) = (-1)^l J_l(k_r r))$, what subsequently leads to a simplified representation of the intensity in the far field [85]:

$$I^{FF}(r, \phi, z) = 2J_l^2(k_r r)((-1^l) + 1 + 2(-1)^l \cos(k_{1z}z - k_{2z}z + 2l\phi)) \quad (2.113)$$

The almost symmetrical structure of OAM Bessel beams becomes obvious here. The number of extrema is modelled by the cosine term and is twice as high as the number of 2π phase jumps. **Shortly after the transition from the near to the intermediate field, a particle that is exposed to this following field, experiences a torque and is thus caused to rotate because the difference of the wave vectors at this point is unequal to zero.**

$$\frac{d\phi}{dz} = \frac{k_{2z} - k_{1z}}{2l} \quad (2.114)$$

As noted at the beginning of the chapter, a common Bessel beam can be realised and calculated with this method, in which only one ring is generated on the SLM. Experimental results and possibilities of both the Bessel and the OAM Bessel beams are investigated in the Sect. 4.8.1.

2.9.4 Laguerre Beams

Laguerre beams denote another beam type covered in this work. The shape of the intensity resembles the already described OAM Bessel beams because this beam type is also referred to as a "doughnut" beam (Fig. 2.25) and also transfers an angular torque. The Laguerre beams do not belong to the non-diffracting beams in a strict sense of this term. However, they are considered as significant due to the wide possibilities in biomedical optics or lithography.

A Laguerre beam is characterised by means of two indices, the azimuthal and the radial one [86]. Such rays are especially helpful for trapping non-transparent particles or particles with a high refractive index.

For generating a Laguerre beam experimentally, a hologram is generated that models the phase of the incident beam radially analogous to the Bessel beams. This

l=2, p=2   l=4, p=1   l=3, p=1

Fig. 2.25 Phase masks modulo 2π for the generation of Laguerre-beams having different azimuth (p), and radial (l) parameters, for instance for display on a spatial light modulator (SLM)

beam type is particularly vulnerable to small pixel errors of the SLM during the realisation of higher order beams. The hologram is illuminated with the inner portion of the Gaussian beam so that the data can be considered as a plane wave [87, 88]. The field distribution of the beam is the next possible solution of the Helmholtz equation (2.87) in cylindrical coordinates. The following complex notation illustrates the structure of the beam [86]:

$$u_p^l(r, \phi, z) = \frac{(-1)^p}{w_z} \left[\frac{2}{\pi} \frac{p!}{(p+|l|)!} \right]^{1/2} \left(\frac{\sqrt{2}r}{w_z} \right)^{|l|}$$

$$\exp\left(-\frac{r^2}{w_z^2}\right) L_p^{|l|} \exp(-il\phi) \exp\left(-i\frac{r^2}{w_z^2}\frac{z}{z_r}\right) \times \exp[i(2p+|l|+1)\arctan(z/z_r)]$$

$$(2.115)$$

In this way the resemblance to the Gaussian beam becomes more obvious, with the constants z_r for the Rayleigh length, $w(z)$ for the beam radius r and k is the wave number of the light. The equation is mainly modulated by the exponential term of the phase shift $\exp(il\phi)$ and the Laguerre polynomial $L_p^{|l|}$. The phase shift caused by the exponential term produces a helical shape of the beam and the Laguerre polynomials are responsible for the rings and their position respectively.

The polynomials have the following form [86, 89]:

$$L_n(x) := \frac{e^x}{n!} \frac{d^n}{dx^n}(x^n e^{-x}) L_n^k(x) = (-1)^k \frac{d^k}{dx^k} L_{n+k}(x) \qquad (2.116)$$

When considering the solution in Cartesian coordinates, the number of signs changes corresponding to the p-values. The required phase pattern on the SLM has a helical structure with an additionally discontinuity and is generated by the following formula [86]:

$$\varphi(r, \phi) = -l\phi + \pi\Theta(L_p^{|l|}(2r^2/w_0^2)), \qquad (2.117)$$

with the Heaviside distribution $\Theta(x)$.

2.9.5 Mathieu Beams

The previously presented beam types are solutions of the Helmholtz equation in Cartesian or cylindrical coordinates. In this section a further beam geometry is introduced denoted as Mathieu beams (Fig. 2.26), that are a solution of the wave equation in elliptical coordinates [90, 91] and are also tested experimentally in this work regarding applicability in a trapping system.

This beam geometry is a further development of the Durnin-approach and presents the wave equation as the Mathieu's differential equation.

Mathieu even 4. order Mathieu odd 8. order   Mathieu odd 4. order Mathieu even 8. order

Fig. 2.26 Various Mathieu phase masks modulo 2π of 4th and 8th order with odd (left) and even (right) parity, for instance for display on a spatial light modulator (SLM)

The form of the wave equation in Cartesian coordinates can be written as follows:

$$\left(\frac{\delta^2}{\delta x^2} + \frac{\delta^2}{\delta y^2} + \frac{\delta^2}{\delta z^2} + k^2\right) E(x, y, z) = 0 \qquad (2.118)$$

By using elliptical coordinates with $x = 1/2d \cosh(u)\cos(v)$, $y = 1/2d \cosh(u)\cos(v)$ and $z = z$, where d is the focal length of an ellipse, a separation of variables in the solution is achieved [90]:

$$W = Z(z)F(u, v) = Z(z)f(u)g(v) \qquad (2.119)$$

Inserting the approach in the wave equation leads to:

$$-\frac{1}{Z}\frac{\delta^2 Z}{\delta z^2} = k_z^2 = k^2 + \frac{1}{F}\left(\frac{\delta^2}{\delta x^2} + \frac{\delta^2}{\delta y^2}\right) F , \qquad (2.120)$$

with the separation constant k_z^2. The general solution for the left part of the equation

$$\left(\frac{\delta^2}{\delta z^2} + k_z^2\right) Z(z) = 0 \qquad (2.121)$$

is given by:

$$Z(z) = \exp(\pm ik_z z). \qquad (2.122)$$

The right part of the equation, however, may be transformed as follows:

$$\left(\frac{\delta^2}{\delta x^2} + \frac{\delta^2}{\delta y^2} + k_t^2\right) [f(u)g(v)] = 0 , \qquad (2.123)$$

with $k_t^2 = k_2 - k_z^2$. By using the two-dimensional Laplace operator

$$\left(\frac{\delta^2}{\delta x^2} + \frac{\delta^2}{\delta y^2}\right) = \frac{1}{\Lambda^2}\left(\frac{\delta^2}{\delta u^2} + \frac{\delta^2}{\delta v^2}\right) \qquad (2.124)$$

in elliptic coordinates, the equation can be transcribed to:

$$a = \frac{f''(u)}{f(u)} + \left(\frac{1}{2}d\right)^2 \frac{1}{2}k_t^2 \cosh(2u) = -\frac{g''(v)}{g(v)} + \left(\frac{1}{2}d\right)^2 \frac{1}{2}k_t^2 \cos(2u) , \quad (2.125)$$

with another separation constant a. This leads to Mathieu's differential equation in the form [90]:

$$\frac{d^2 f(u)}{du^2} - (a - 2q\cosh(2u))f(u) = 0 \qquad (2.126)$$

and

$$\frac{d^2 g(v)}{dv^2} - (a - 2q\cos(2v))g(v) = 0 \qquad (2.127)$$

with $q = \frac{1}{4}k_t^2(\frac{1}{2}d)^2$. The function (2.127) is $n\pi$-periodical, but only π and 2π-periodic solutions are considered, since it corresponds to the wave equation. Furthermore, it can be shown that a lot of even or odd solutions for (2.127) exist, which depend on the characteristic values given in the following.

To generate the Mathieu geometry, a corresponding phase mask is displayed on the SLM. The even or odd solutions and the order of the beam are determining the shape of the beam. The mask is generated according to the Cojacaru procedure [91] and is given by

$$M_r^e(u, v) = sgn\{ce_r(v, q)Je_r(u, q)\} \qquad (2.128)$$

for the even mode and

$$M_r^o(u, v) = sgn\{se_r(v, q)Jo_r(u, q)\} \qquad (2.129)$$

for the odd mode, with the common Mathieu functions ce_r and se_r and the so called modified Mathieu functions Je_r and Jo_r. The spatial coordinate is rotated by $\pm 90°$ in the modified functions: $z \to \pm iz$.

2.9.6   Airy Beams

One of the latest beam geometries studied in research is the Airy beam type (see Fig. 2.27). Its propagation characteristics are also diffractive, but the beam contains a transverse acceleration which results in a parabolic trajectory of the beam. This effect provides applications of this beams in cell biology, since cell transport along the just mentioned paths can be realised with relatively little effort [92]. Therefore, also this beam type is investigated concerning applicability in a developed trapping system.

The intensity patterns typically consist of a main maximum and several secondary maxima, which are arranged at a 90 angle to each other and are relatively close to

Fig. 2.27 Phase mask modulo 2π for creating an Airy beam

the main maximum. The gradient force in direction to an increased intensity drag the micro-particles towards the main maximum.

At the maximum of the parabolic trajectory there is a critical point at which the forces are weakest and the beam looses its partial non-diffractive character. Consequently, the gradient force is relatively weak here, causing that the trapping force of the beam is no longer sufficient in this region and the trapped particles move then down to a new position.

The Airy beams also arise as a solution of the Helmholtz equation (see also Appendix B.14) in the paraxial representation [92]:

$$\delta_x^2 u_o(x, y, z) + \delta_y^2 u_o(x, y, z) - 2ik\delta_z^2 u_o(x, y, z) = 0 \qquad (2.130)$$

with the normalisation vector $k = 2\pi n/\lambda$ in the direction of propagation z. The wave has the form $\exp(i(\omega t - kz))$ and $u_o(x, y, z)$ is the scalar field, which can be divided in the electric $\mathbf{E}$ and magnetic field $\mathbf{H}$, with the help of the Lorentz vector potential $\mathbf{A}$:

$$\mathbf{A} = u_o(x, y, z)\exp(-ikz)\vec{\mathbf{x}}$$

$$\mathbf{E} = -ik\mathbf{A} - \frac{i}{k}\nabla(\nabla\mathbf{A}) \qquad (2.131)$$

$$\mathbf{H} = \sqrt{\frac{\epsilon_0}{\mu_0}}\nabla\times\mathbf{A}$$

Here the fields in x-direction are considered, μ_0 and ϵ_0 are constants of space and $\exp(i\omega t)$ the time dependence of the wave is omitted for the sake of clarity. In the solution considered here, the separation method which was also used in the previous section, is employed for the scalar field $u_0(x, y, z) = u_x(x, y)u_y(y, z)$:

$$0 = -2ik\delta_z u_x + \delta_x^2 u_x$$
$$0 = -2ik\delta_z u_y + \delta_y^2 u_y \qquad (2.132)$$

Here the x direction will be regarded, the y-direction can be calculated similarly. The scalar field in the maximum of the parabola ($z = 0$, parabola $z = b_o x^2$) of the Airy beam is specified by

$$u_x(x, z = 0) = Ai\left(\frac{x}{x_0}\right) \exp\left(\frac{a_0 x}{x_0}\right) , \qquad (2.133)$$

with the Airy function Ai, the characteristic length x_0 and the coefficient of the aperture a_0. According to the principle of holography, the desired field in the focal plane corresponds to the Fourier transform of the reflected beam from the SLM. Consequently, the field at the SLM is given by [92]

$$u_x(k, z = z_{\text{SLM}}) = x_0 \exp(-a_0 x_0^2 k2) \exp\left(\frac{i}{3}(x_0^3 k^3 - 3a_0^2 x_0 k - ia_0^3)\right). \qquad (2.134)$$

The equation can be divided into two parts. The first part corresponds in this case to the incident Gaussian beam which originates from the laser, and the second part describes the phase mask loaded into the SLM.

A more accurate and flexible solution of the (2.132) is given by the Huygens–Fresnel integral. This integral allows to write the scalar field u_x in dependence of z and the value of the scalar field in $z = 0$:

$$u_x(x, z) \quad = \int \sqrt{\frac{ik}{2\pi z}} \exp\left(-\frac{k}{2z}(x_1^2 - 2x_1 x + x^2)\right) u_x(x, z = 0) \qquad (2.135)$$

$$= \int \sqrt{\frac{ik}{2\pi z}} \exp\left(-\frac{k}{2z}(x_1^2 - 2x_1 x + x^2)\right) Ai\left(\frac{x_1}{x_0}\right) e^{a_0 x_1/x_0} dx_1 \qquad (2.136)$$

$$= Ai\left(\frac{x - x_m(z)}{x_0} + \frac{a_0 z}{k x_0^2}\right) \exp\left(\frac{a_0(x - x_m(z))}{x_0} - i\frac{z^3}{12k^3 x_0^6} + i\frac{zx}{2kx_0^3}\right) \qquad (2.137)$$

with $x_m(z) = z2/(4k^2 x_0^3)$ [92].

2.10 Direct Laser Writing with Two-Photon Polymerisation

In contrast to one-photon absorption (OPA), two-photon absorption (TPA) is a non-linear process where two photons are absorbed simultaneously by e.g. an atom or molecule. This was first described in 1931 by Maria Goeppert-Mayer [93]. One of these photons alone does not have enough energy to reach the excited state.

However, since 2PP occurs only in the rarest cases under normal conditions, the process assumes a very high temporal and spatial photon density. For this reason, the 2PP could only be realized shortly after the invention of the laser. DLW with high resolution is only possible for laser wavelengths in which one-photon absorption is strongly suppressed in the photoresist and two-photon absorption is preferred. This is the case when the resist becomes transparent to the wavelength of the laser λ and simultaneously absorbing at the wavelength $\lambda/2$. Two photons can excite an atom to a higher energy state whereat the energy increase is given by the sum of the two photon energies. The effect of TPA was proposed by M. Goeppert-Mayer in 1931 [94]. But only with the invention of the laser by T. Maiman in 1960, an application of this phenomenon became possible. In 1961, W. Kaiser and C.G.B. Garrett were the first to experimentally observe the phenomenon of TPA [95]. In order to explain the mechanism of TPA, an energy system of a molecule is assumed having a ground, an intermediate virtual and an upper state. The Jablonski-diagram in Fig. 2.28 illustrates the processes of OPA, TPA and fluorescence. For an OPA, the photon-energy must be equal to the energy difference of the ground state and a vibrational state of the upper state. For TPA, only the combined energy of two photons is sufficient to reach the upper state. In contrast to a single photon absorption, TPA requires a high spatial as well as a sufficient temporal overlap of the incident photons, because the virtual state has a vanishing short life-time. Additionally, there exist selection rules for a TPA which are the exact opposite compared to those for a OPA [96]. Therefore, TPA can induce an excitation process which can be forbidden for OPA. The last depicted process in Fig. 2.28 is called fluorescence. After nonradiative transitions to the lowest vibrational state of the excited state, the system can relax to the ground state upon the emission of a photon with appropriate energy. Regarding TPA, high light intensities are necessary to increase the probability of such an event. A major reason why it took 30 years from the theoretical invention of TPA to the first experimental realisation was the availability of high light intensities which became technically feasible by the invention of the laser. By regarding a second-order perturbation theory, a rate equation describing the TPA probability can be deduced [97]:

$$\frac{dN}{dt} = \frac{\bar{\sigma} C V}{(hf)^2} \cdot I(z, t)^2 \ . \tag{2.138}$$

Here, $\bar{\sigma}$ is the TPA cross section, C is the concentration of the absorber, V is the reaction volume, f is frequency and $I(z, t)$ denotes the instantaneous intensity of the exciting light. The second power of the intensity denotes the non-linearity of the process. For a further increase of the intensity, spatial confinement as well as a specific time modulation of the laser beam can be used. By focusing the light for instance with a lens, the intensities in the vicinity of the focal point can be increased. Femtosecond laser pulses can be used for a further improvement. Their advantage is, that high peak intensities can be realised without transferring too much energy in time average. TPA is the basis for two-photon polymerisation (TPP) which means a chemical reaction of molecules when forming polymer chains by a simultaneous absorption of two photons.

Fig. 2.28 Jablonski-diagram illustrating the TPA process

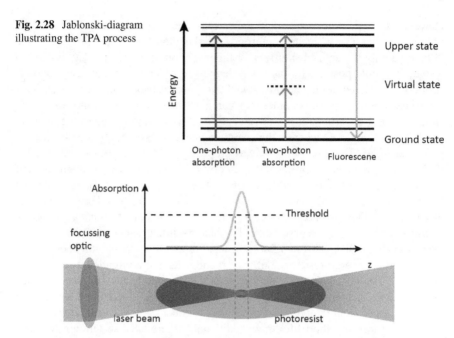

Fig. 2.29 Focusing of light in order to meet the absorption threshold required for TPP

When the intensity rises, TPA becomes an increasingly likely event in a suitable photoresist. When the absorption in the resist meets a certain threshold value, the process of TPP is initiated. This is visualised in Fig. 2.29. The exposed material in which the process of TPP is triggered undergoes a significant change of physical properties such as the solubility respective to specific solvents. Thus, the process of TPP can be used for lithographic means [98, 99].

References

1. Gabor, D.: A new microscopic principle. Nature **161**, 777–778 (1948)
2. Menzel, E., Mirandé, W., Weingärtner, I.: Fourier-Optik und Holographie, vol. 1, pp. 140–142. Springer, Wien (1973)
3. Kreis, T.: Handbook of Holographic Interferometry: Optical and Digital Methods. Wiley-VCH, Weinheim (2005). ISBN 9783527405466
4. Poon, T.-C.: Digital Holography and Three-Dimensional Display: Principles and Applications, vol. 1. Springer, Wien (2006)
5. Yaroslavsky, L.: Introduction to Digital Holography [Saif Zone and Sharjah and U.A.E.], vol. 1. Bentham eBooks (2009). ISBN 9781608050796
6. Kreis, T.: Holographic Interferometry: Principles and Methods. Akademie Verlag Series in Optical Metrology. Akademie Verlag (1996). http://books.google.de/books? id=qfJRAAAAMAAJ, ISBN 9783055016448

7. Leith, E.N., Upatnieks, J.: Reconstructed wavefronts and communication theory. J. Opt. Soc. Am. **52**(10), 1123–1128 (1962). https://doi.org/10.1364/JOSA.52.001123

8. Leith, E.N., Upatnieks, J.: Wavefront reconstruction with continuous-tone objects. J. Opt. Soc. Am. **53**(12), 1377–1381 (1963). https://doi.org/10.1364/JOSA.53.001377

9. Goodman, J.W.: Introduction to Fourier Optics. McGraw-Hill Physical and Quantum Electronics Series. Roberts & Company (2005). http://books.google.de/books?id=ow5xs_Rtt9AC, ISBN 9780974707723

10. Lesem, L.B., Hirsch, P.M.: The kinoform: a new wavefront reconstruction device. IBM J. Res. Dev. **13**, 150–155 (1969)

11. Schnars, U., Jueptner, W.: Direct recording of holograms by a CCD target and numerical reconstruction. Appl. Opt. **33**(2), 179–181 (1994). https://doi.org/10.1364/AO.33.000179

12. Schnars, Ulf., Jüptner, W.P.O.: Digital recording and numerical reconstruction of holograms. Meas. Sci. Technol. **13**(9), R85 (2002). http://stacks.iop.org/0957-0233/13/i=9/a=201

13. Schnars, Ulf., Jüptner, W.: Digital Holography. Springer, Berlin (2005)

14. Gerchberg, R., Saxton, W.: A practical algorithm for the determination of phase from image and diffraction plane pictures. Optik **35**(2), 237–246 (1972)

15. Allebach, J., Seldowitz, M.: Synthesis of digital holograms by direct binary search. Appl. Opt. **26**(14), 2788–2798 (1987)

16. Schmit, J., Creath, K.: Extended averaging technique for derivation of error-compensating algorithms in phase-shifting interferometry. Appl. Opt. **34**(19), 3610–3619 (1995). https://doi.org/10.1364/AO.34.003610

17. El Jarad, A., Gulker, G., Hinsch, K.D.: Microscopic ESPI: better fringe qualities by the Fourier transform method. Proc. SPIE **4933**, 335–341 (2003). https://doi.org/10.1117/12.516662

18. Mills, G.A., Yamaguchi, I.: Effects of quantization in phase-shifting digital holography. Appl. Opt. **44**(7), 1216–1225 (2005). https://doi.org/10.1364/AO.44.001216

19. Schwider, J., Burow, R., Elssner, K.-E., Grzanna, J., Spolaczyk, R., Merkel, K.: Digital wavefront measuring interferometry: some systematic error sources. Appl. Opt. **22**(21), 3421–3432 (1983). https://doi.org/10.1364/AO.22.003421

20. Malacara, D.: Optical Shop Testing, vol. 3. Wiley-VCH, New Jersey (2007)

21. Malacara, D.: Optical Shop Testing. Wiley Series in Pure and Applied Optics. Wiley, New York (2007). http://books.google.de/books?id=qMHKB1mKFr4C, ISBN 9780470135969

22. Burke, J., Helmers, H.: Spatial versus temporal phase shifting in electronic speckle-pattern interferometry: noise comparison in phase maps. Appl. Opt. **39**(25), 4598–4606 (2000). https://doi.org/10.1364/AO.39.004598

23. Brophy, C.P.: Effect of intensity error correlation on the computed phase of phase-shifting interferometry. J. Opt. Soc. Am. A **7**(4), 537–541 (1990). https://doi.org/10.1364/JOSAA.7.000537

24. Bothe, T., Burke, J., Helmers, H.: Spatial phase shifting in electronic speckle pattern interferometry: minimization of phase reconstruction errors. Appl. Opt. **36**(22), 5310–5316 (1997). https://doi.org/10.1364/AO.36.005310

25. Liebling, M., Blu, T., Cuche, E., Marquet, P., Depeursinge, C., Unser, M.: A novel non-diffractive reconstruction method for digital holographic microscopy. In: Proceedings of the 2002 IEEE International Symposium on Biomedical Imaging, pp. 625–628 (2002)

26. Liebling, M., Blu, T., Unser, M.: Complex-wave retrieval from a single off-axis hologram. J. Opt. Soc. Am. A **21**(3), 367–377 (2004). https://doi.org/10.1364/JOSAA.21.000367

27. Carl, D., Kemper, B., Wernicke, G., von Bally, G.: Parameter-optimized digital holographic microscope for high-resolution living-cell analysis. Appl. Opt. **43**(36), 6536–6544 (2004). https://doi.org/10.1364/AO.43.006536

28. Stuerwald, S., Schmitt, R.: Readjusting image sharpness by numerical parametric lenses in Forbes-representation and Halton sampling for selective refocusing in digital holographic microscopy - Errata. Version 2010. https://doi.org/10.1117/12.903693

29. Remmersmann, C., Stürwald, S., Kemper, B., Langehanenberg, P., von Bally, G.: Phase noise optimization in temporal phase-shifting digital holography with partial coherence light sources and its application in quantitative cell imaging. Appl. Opt. **48**(8), 1463–1472 (2009). https://doi.org/10.1364/AO.48.001463

30. Cuche, E., Depeursinge, C.: Digital holography for quantitative phase-contrast imaging. Opt. Lett. **24**(5), 291–293 (1999)
31. Marquet, P., Rappaz, B., Magistretti, P.J., Cuche, E., Emery, Y., Colomb, T., Depeursinge, C.: Digital holographic microscopy: a noninvasive contrast imaging technique allowing quantitative visualization of living cells with subwavelength axial accuracy. Opt. Lett. **30**(5), 468–470 (2005). https://doi.org/10.1364/OL.30.000468
32. Dubois, F., Requena, M.-L.N., Minetti, C., Monnom, O., Istasse, E.: Partial spatial coherence effects in digital holographic microscopy with a laser source. Appl. Opt. **43**(5), 1131–1139 (2004). https://doi.org/10.1364/AO.43.001131
33. Shamir, J.: Press Monographs. Optical Systems and Processes, vol. 3. OSA (1986). https://doi.org/10.1364/JOSAA.3.000847, ISBN 9780819432261
34. Lankenau, E., Klinger, D., Winter, C., Malik, A., Müller, H., Oelckers, S., Pau, H.-W., Just, T., Hüttmann, G.: Combining optical coherence tomography (OCT) with an operating microscope. In: Buzug, T.M., Holz, D., Bongartz, J., Kohl-Bareis, M., Hartmann, U., Weber, S. (eds.) Advances in Medical Engineering, vol. 114, pp. 343–348. Springer, Berlin (2007). ISBN 978–3–540–68763–4
35. Haferkorn, H.: Optik: Physikalisch-technische Grundlagen und Anwendungen, vol. 4. Wiley-VCH Verlag (2002)
36. Jahns, J.: Photonik: Grundlagen, Komponenten und Systeme. Oldenbourg Wissenschaftsverlag (2000)
37. vision.at/images01/DMD2.JPG, 14.08.2012. InVision. http://www.science-vision.at/images01/DMD2.JPG
38. Liesener, J.: Zum Einsatz räumlicher Lichtmodulatoren in der interferometrischen Wellenfrontmesstechnik. Dissertation, Universität Stuttgart, Stuttgart, 24.03.2007
39. Texas Instruments: Introduction to digital micromirror device (DMD) technology (2008). http://www.ti.com/lit/an/dlpa008/dlpa008.pdf
40. Fraunhofer IPMS: MEMS phase former kit. http://www.ipms.fraunhofer.de/content/dam/ipms/common/products/SLM/phase-former-kit-e.pdf
41. Dai, H., Liu, K.X.Y., Wang, X., Liu, J.: Characteristics of LCoS phase-only spatial light modulator and its applications. Opt. Commun. **238**(4–6), 269–276 (2004). ISSN 00304018
42. Hecht, E.: Optik, vol. 5. Oldenbourg, München (2009). ISBN 9783486588613
43. Wilkinson, T.D., Henderson, C.D., Leyva, D.G., Crossland, W.A.: Phase modulation with the next generation of liquid crystal over silicon technology. J. Mater. Chem. **16**(33), 3359 (2006). ISSN 0959–9428
44. Serati, S., Xia, X.: High-resolution phase-only spatial light modulators with submillisecond response. In: SPIE Proceedings, vol. 5106 (2003)
45. Serati, S., Harriman, J.: Spatial light modulator considerations for beam control in optical manipulation applications. Version 2006. https://doi.org/10.1117/12.681156
46. Lizana, A., Márquez, A., Lobato, L., Rodange, Y., Moreno, I., Iemmi, C., Campos, J.: The minimum Euclidean distance principle applied to improve the modulation diffraction efficiency in digitally controlled spatial light modulators. Opt. Express **18**(10), 10581–10593 (2010). https://doi.org/10.1364/OE.18.010581
47. Kohler, C.: Optimierung von Flüssigkristall-Lichtmodulatoren in aktiven optischen Systemen. Dissertation, Universität Stuttgart, Stuttgart, 23.07.2009
48. Zwick, S.: Flexible Mikromanipulation durch räumliche Lichtmodulation in der Mikroskopie. Dissertation, Universität Stuttgart, Stuttgart, 19.05.2010
49. Ashkin, A.: Acceleration and trapping of particles by radiation pressure. Phys. Rev. Lett. **24**(4), 156–159 (1970)
50. Maurer, C., Jesacher, A., Bernet, S., Ritsch-Marte, M.: What spatial light modulators can do for optical microscopy. Laser Photonics Rev. **5**(1), 81–101 (2011). https://doi.org/10.1002/lpor.200900047, ISSN 18638880
51. Ashkin, A.: History of optical trapping and manipulation of small neutral particles, atoms, and molecules. Springer Series in Chemical Physics, vol. 67, pp. 1–31. Springer, Berlin (2001). https://doi.org/10.1007/978-3-642-56544-1_1, ISBN 978–3–642–62702–6

52. Svoboda, K., Block, S.M.: Biological applications of optical forces. Annu. Rev. Biophys. Biomol. Struct. **23**(1), 247–285 (1994). https://doi.org/10.1146/annurev.bb.23.060194.001335, PMID: 7919782

53. Fällman, E., Axner, O.: Design for fully steerable dual-trap optical tweezers. Appl. Opt. **36**(10), 2107–2113 (1997). https://doi.org/10.1364/AO.36.002107

54. Sasaki, K., Koshioka, M., Misawa, H., Kitamura, N., Masuhara, H.: Pattern formation and flow control of fine particles by laser-scanning micromanipulation. Opt. Lett. **16**(19), 1463–1465 (1991). https://doi.org/10.1364/OL.16.001463

55. Brouhard, G.J., Schek, H.J., Hunt, A.J.: Advanced optical tweezers for the study of cellular and molecular biomechanics. IEEE Trans. Biomed. Eng. **50**(1), 121–125 (2003). https://doi.org/10.1109/TBME.2002.805463

56. Dufresne, E.R., Spalding, G.C., Dearing, M.T., Sheets, S.A., Grier, D.G.: Computer-generated holographic optical tweezer arrays. Rev. Sci. Instrum. **72**(3), 1810–1816 (2001). https://doi.org/10.1063/1.1344176

57. Dufresne, E.R., Grier, D.G.: Optical tweezer arrays and optical substrates created with diffractive optics. Rev. Sci. Instrum. **69**(5), 1974–1977 (1998). https://doi.org/10.1063/1.1148883

58. Reicherter, M., Haist, T., Wagemann, E.U., Tiziani, H.J.: Optical particle trapping with computer-generated holograms written on a liquid-crystal display. Opt. Lett. **24**(9), 608–610 (1999). https://doi.org/10.1364/OL.24.000608

59. Curtis, J.E., Koss, B.A., Grier, D.G.: Dynamic holographic optical tweezers. Opt. Commun. **207**(1–6), 169–175 (2002). https://doi.org/10.1016/S0030-4018(02)01524-9, ISSN 0030-4018

60. Kawata, Y., Fujita, K.: 4Pi confocal optical system with phase conjugation. Opt. Lett. **21**(18), 1415–1417 (1996)

61. Knox, K., Burnham, D.: Observation of bistability of trapping position in aerosol optical tweezers. J. Opt. Soc. Am. B **27**(3), 582–591 (2010)

62. Ashkin, A.: Optical trapping and manipulation of neutral particles using lasers. Proc. Natl. Acad. Sci. USA **94**, 4853–4860 (1997)

63. Ashkin, A., Dziedzic, J.M., Bjorkholm, J.E., Chu, S.: Observation of a single-beam gradient force optical trap for dielectric particles. Opt. Lett. **11**(5), 288–290 (1986). https://doi.org/10.1364/OL.11.000288

64. Ashkin, A.: Forces of a single-beam gradient laser trap on a dielectric sphere in the ray optics regime. Biophys. J. **61**, 569–582 (1992)

65. Svoboda, K., Block, S.M.: Biological applications of optical forces. Dissertation, Harvard University, Cambridge and Massachusetts (1994)

66. Reicherter, M.: Optical particle trapping with computer-generated holograms written on a liquid-crystal display. Opt. Lett. **24**(9), 608–610 (1999)

67. Wright, W.H., Sonek, G.J., Berns, M.W.: Parametric study of the forces on microspheres held by optical tweezers. Appl. Opt. **33**(9), 1735–1748 (1994)

68. Wright, W.H.: Radiation trapping forces on microspheres with optical tweezers. Appl. Phys. Lett. **63**(6), 715–717 (1993)

69. Hwang, S.-Uk., Park, Y.-H., Lee, Y.-G.: Interactive Control of holographic optical traps with fast hologram generation. IEEE **1**, 183–188 (2009)

70. Saleh, B.E.A., Teich, M.C.: Grundlagen der Photonik. Lehrbuch Physik, vol. 2, vollst. überarb. und erw. Aufl. [=1. dt. Aufl.]. Wiley-VCH, Weinheim (2008). ISBN 9783527406777

71. Reicherter, M.: Einsatz von Lichtmodulatoren zum Teilcheneinfang und zur Aberrationskontrolle in holografischen Pinzetten. Dissertation, Universität Stuttgart, Stuttgart, 29.09.2006

72. Kegler, K., Salomo, M., Kremer, F.: Forces of interaction between DNA-grafted colloids: an optical tweezer measurement. Phys. Rev. Lett. **98**, 058304 (2007). https://doi.org/10.1103/PhysRevLett.98.058304

73. Perkins, T.T.: Optical traps for single molecule biophysics: a primer. Laser Photonics Rev. **3**(1–2), 203–220 (2009). ISSN 18638880

74. Enger, J.: Optical tweezers applied to a microfluidic system. Lab Chip **4**(3), 196–200 (2004). ISSN 1473–0197

75. Seeger, S.: Application of laser optical tweezers in immunology and molecular genetics. Cytometry **12**, 497–504 (1991)

76. Ladavac, K., Grier, D.: Microoptomechanical pumps assembled and driven by holographic optical vortex arrays. Opt. Express **12**(6), 1144–1149 (2004)

77. Jackson, J.D.: Klassische Elektrodynamik. de Gruyter (1981). http://books.google.de/books?id=JFdwygAACAAJ, ISBN 9783110074154

78. Durnin, J.: Exact solutions for nondiffracting beams. I. The scalar theory. J. Opt. Soc. Am. A **4**(4), 651–654 (1987). https://doi.org/10.1364/JOSAA.4.000651

79. Arlt, J., Garces-Chavez, V., Sibbett, W., Dholakia, K.: Optical micromanipulation using a Bessel light beam. Opt. Commun. **197**(4–6), 239–245 (2001). https://doi.org/10.1016/S0030-4018(01)01479-1, ISSN 0030–4018

80. Born, M., Wolf, E., Bhatia, A.B.: Principles of Optics: Electromagnetic Theory of Propagation, Interference and Diffraction of Light. Cambridge University Press, Cambridge (1999). http://books.google.de/books?id=aoX0gYLuENoC, ISBN 9780521642224

81. Wright, E.M.: The generalized Bessel function of order greater than one. Q. J. Math. **os-11**(1), 36–48 (1940). https://doi.org/10.1093/qmath/os-11.1.36

82. Gori, F., Guattari, G., Padovani, C.: Bessel–Gauss beams. Opt. Commun. **64**(6), 491–495 (1987). https://doi.org/10.1016/0030-4018(87)90276-8, ISSN 0030–4018

83. Arlt, J., Dholakia, K.: Generation of high-order Bessel beams by use of an axicon. Opt. Commun. **177**(1–6), 297–301 (2000). https://doi.org/10.1016/S0030-4018(00)00572-1, ISSN 0030–4018

84. Vasara, A., Turunen, J., Friberg, A.T.: Realization of general nondiffracting beams with computer-generated holograms. J. Opt. Soc. Am. A **6**(11), 1748–1754 (1989). https://doi.org/10.1364/JOSAA.6.001748

85. Chattrapiban, N., Rogers, E.A., Cofield, D., Hill III, W.T., Roy, R.: Generation of nondiffracting Bessel beams by use of a spatial light modulator. Opt. Lett. **28**(22), 2183–2185 (2003). https://doi.org/10.1364/OL.28.002183

86. Allen, L., Beijersbergen, M.W., Spreeuw, R.J.C., Woerdman, J.P.: Orbital angular momentum of light and the transformation of Laguerre–Gaussian laser modes. Phys. Rev. A **45**, 8185–8189 (1992). https://doi.org/10.1103/PhysRevA.45.8185

87. Matsumoto, N., Ando, T., Inoue, T., Ohtake, Y., Fukuchi, N., Hara, T.: Generation of high-quality higher-order Laguerre–Gaussian beams using liquid-crystal-on-silicon spatial light modulators. J. Opt. Soc. Am. A **25**(7), 1642–1651 (2008). https://doi.org/10.1364/JOSAA.25.001642

88. Debailleul, M., Simon, B., Georges, V., Haeberl, O., Lauer, V.: Holographic microscopy and diffractive microtomography of transparent samples. Meas. Sci. Technol. **19**(7), 074009 (2008). http://stacks.iop.org/0957-0233/19/i=7/a=074009

89. Kimel, I., Elias, L.R.: Relations between Hermite and Laguerre Gaussian modes. IEEE J. Quantum Electron. **29**(9), 2562–2567 (1993). https://doi.org/10.1109/3.247715, ISSN 0018–9197

90. Stamnes, J.J., Spjelkavik, B.: New method for computing eigenfunctions (Mathieu functions) for scattering by elliptical cylinders. Pure Appl. Opt. J. Eur. Opt. Soc. Part A **4**(3), 251 (1995). http://stacks.iop.org/0963-9659/4/i=3/a=011

91. Cojocaru, E.: Mathieu functions computational toolbox implemented in Matlab (2008). arXiv:0811.1970, Forschungsbericht, Comments: 20 pages, 0 figures, 6 tables

92. Morris, J.E., Mazilu, M., Baumgartl, J., Cizmar, T., Dholakia, K.: Supercontinuum Airy beams. Version: 2009. https://doi.org/10.1117/12.826098

93. Goeppert-Mayer, M.: Ueber Elementarakte mit zwei Quantenspruengen. Dissertation, Universitaet Goettingen (1931)

94. Goeppert-Mayer, M.: Ueber Elementarakte mit zwei Quantenspruengen. Annalen der Physik **401**(3), 273–294 (1931). https://doi.org/10.1002/andp.19314010303, ISSN 1521–3889

95. Bayer, E., Schaack, G.: Two-photon absorption of CaF2:Eu2+. Phys. Status Solidi (b) **41**(2), 827–835 (1970). https://doi.org/10.1002/pssb.19700410239, ISSN 1521–3951

96. McClain, M.: Two-photon molecular spectroscopy. Phys. Rev. Lett. (1974)

97. Kafri, O., Kimel, S.: Theory of two-photon absorption and emission second-order saturation effect. Chem. Phys. **5**(3), 488–493 (1974). https://doi.org/10.1016/0301-0104(74)85052-4, ISSN 0301–0104
98. Nanoscribe GmbH. http://www.nanoscribe.de/
99. Schaeffer, S.: Characterization of two-photon induced cross-linking of proteins. Master thesis, RWTH Aachen (2013)

Chapter 3
Available Systems and State of the Art

This section gives a brief overview of the state of the art of (commercially) available systems for digital holographic microscopy (DHM) and quantitative phase contrast, 3D-nanopositioning, optical tweezers as well as 3-dimensional lithography by direct laser writing. Those systems, that are utilized for the experimental setups in the following chapters, are introduced in more detail.

3.1 Systems for Optical Traps

The first modules for optical tweezers, with which conventional light microscopes could be upgraded, were based on a highly focused laser beam allowing the trapping of individual micro-particles in the focus area. In order to position this particle, the microscope stage is moved, while the laser beam along with the "caught" particle remains rigid (e.g. E3300 from Elliot Scientific [1]). One disadvantage of these systems is the inability to trap a plurality of particles, and to move them relative to each other which is required for many biomedical applications.

One approach to produce several trap spots is the use of a split beam with partial beams that can be moved laterally by means of scanner systems each consisting of two galvo mirrors. However, with an increasing number of traps, such systems become relatively complex, slow and error-prone as the deflecting mechanical elements are controlled independently (e.g. NanoTracker of JPK Instruments [2]).

The manufacture of many optical traps without using mechanical moving parts can be based on diffractive optical elements (DOE) that deflect most of the light in the desired diffraction order. Acousto-optical deflectors, also called Bragg cells, represent such DOE. Inducing ultrasonic in a transparent crystal body creates a standing wave. This acts as a diffraction grating that deflects the transmitted laser beam [3]. By adjusting the sound frequency and thus the lattice constant, the angle of diffraction of the beam can be changed. Since no mechanical effects are used,

© Springer Nature Switzerland AG 2018
S. Stuerwald, *Digital Holographic Methods*, Springer Series
in Optical Sciences 221, https://doi.org/10.1007/978-3-030-00169-8_3

high switching times in the range of a few microseconds are possible. The limiting factor here is the speed of sound in the crystal in use. With this procedure several time shared traps can be generated by scanning the corresponding positions with fast multiplexing of different frequencies. As the medium, including particles, becomes more inert at higher scanning frequencies, a relatively high number of micro-particles can be trapped without remark of a sequential movement. However, for many spots (approx. > 100) or complex geometries, this method also has its limitations, as the time intervals between exposures of the same spots may become too long to ensure a stable trap [4, pp. 122ff]. For this procedure, upgrade modules are offered for conventional light microscopes (e.g. *TWEEZ 200si* from Aresis [5]).

The method applied in this work uses SLMs (Spatial Light Modulators) as DOE in a matrix form, which can in general change the amplitude and/or phase of a transmitted or reflected beam. The utilised micro screens display a diffraction grating or hologram in which the light beam is diffracted - similar to the method described above - to the desired intensity pattern of the object plane. Different designs and types for these SLMs are existing (see Sect. 2.5). The advantage of this variant is that even complex intensity patterns can be generated and displayed on a reflective or transmissive panel without multiplexing. The SLM matrix projects the desired pattern (Fourier hologram) on the amplitude and/or phase distribution of the beam and by means of a projection lens the diffraction pattern is formed. The lens finally performs an optical Fourier transform. Therefore the inverse Fourier transformation of the desired intensity pattern has to be calculated and sent to the SLM in advance.

By outsourcing this computation on the graphics card real-time applications of complex patterns are possible on a standard PC. A major disadvantage of this method is the partition of the laser beam on all generated spots. Thus with increasing number of traps a correspondingly higher laser power is necessary. The company "Arryx" offers an extension package for conventional light microscopes as well as a complete system for digital holographic optical tweezers (HOT-kit, BioRyx 200 [6]), however the utilised system components are no longer state of the art with regard to resolution and speed.

3.2   Imaging by Means of Digital Holographic Quantitative Phase-Contrast Methods

The digital holographic quantitative phase-contrast method for non-invasive imaging of living cells was first introduced in 2004. The advantage in contrast to common interferometric microscope methods such as phase contrast and differential interference contrast (DIC) microscopy, DHM allows a quantitative measurement of parameters such as the phase distribution induced by biologic cells. By numerical reconstruction of the complex wave front from the measured hologram, the intensity and phase distribution can be calculated. In addition, with DHM numerical focusing on different object planes without any opto-mechanical motion is possible [7]. The company "Phase Holographic Imaging" sells special microscopes for

DHM-phase contrast of biological cells (HoloMonitor [8]). In addition, the startup company "Lyncée Tec", emerged from a research group in Lausanne, offers several types of devices each specialized on specific application areas. They are focusing on the development and sale of complete systems for digital holographic microscopy and offer special microscopes for reflection and transmission microscopy [9]. The applied method method is also primarily mentioned in research and elsewhere hardly used and commercialised [10–12].

3.3 Overview of HOT-Systems in Research

The concept of single optical tweezers has been well known for more than twenty years. The immanent intensity gradient of a TEM_{00} laser beam is used to transfer a resulting force to dielectric particles with sizes from tens of nanometres to many micrometres [13]. Since single optical tweezers are relatively easy to implement and powerful in their applications, they have become established as important tools in biological and medical sciences [14]. However, single optical tweezers are limited to the manipulation of one object at a time. Consequently, there have been many concepts developed to allow for manipulation of two or more objects simultaneously. Probably the most obvious approach is taking two laser beams and coupling them into the same optical tweezer setup [15]. If both beams are prepared separately before they are joined, both traps can be controlled independently. This configuration may be called spatial multiplexing. Another approach is time multiplexing of two or more traps [16]. Here, a single laser beam is deflected by a fast pivoting mirror such as a galvano- or piezo-mirror or by acousto-optic devices [17]. The beam is placed at the desired trapping position, kept there for a certain time and moved to the next position. Both concepts are subject to certain restrictions. Spatial multiplexing requires one beam per trap to be prepared and thus the effort scales with the number of traps. As a consequence, most optical tweezers realised on the basis of spatial multiplexing are limited to two independent traps. Temporal multiplexing, on the other hand, requires sharing the trapping time and laser power between all traps. Trap stiffness is thus significantly reduced.

Holographic optical tweezers represent an extraordinary flexible way to create multiple traps [18]. A computer-calculated hologram is placed in the optical path and thereby read out by a reference wave. Commonly the hologram is positioned in a Fourier plane with respect to the trapping plane. The hologram can be designed such that in the trapping plane almost any arbitrary intensity distribution can be achieved. Multiple optical traps in this scenario are only a special case of possible complex trapping geometries. Strong optical tweezers require a high level of laser power in the trapping plane. Consequently, a high diffraction efficiency is mandatory and thus usually phase holograms are used. The required hologram can be produced, for example, by lithographic techniques [18, 19]. A far more flexible way is given by dynamic holographic optical tweezers [20, 21], where the hologram is created by a computer addressable spatial light modulator (SLM). This allows changing trapping geometries without any changes in the optical setup by just giving a new hologram on the SLM. A drawback of computer-generated holograms is that any local change in the trapping geometry requires the calculation of a completely new hologram. Hologram calculation time thus becomes a serious issue in real-time applications.

Still there is a significant drawback with direct imaging methods. As in any tweezers set-up with an SLM, the SLM can be the bottleneck if high trapping force and thus high laser power is required. Direct imaging approaches - as well as holographic optical tweezers - require that the main laser power has to pass through the modulator. The maximal trapping power and the maximal number of traps are limited by the reflection or transmission coefficient, the diffraction efficiency and finally by the damage threshold of the SLM. Though, in literature, first approaches with high laser power and liquid cooled SLM have been demonstrated [22].

3.4 Direct Laser Writing Lithography

Since 2007 the start-up company Nanoscribe from the Karlsruhe Institute of Technology (KIT) offers a commercialised version of a direct laser writing (DLW) research system based on two photon polymerisation. The system is based on a standard inverted microscope (Fig. 3.1) and offers highest achievable resolution of approximately 200 nm × 400 nm voxel size with its adapted microscope objectives. The yellowness in Fig. 3.1 is caused by amber light in order to protect the photoresists from unwanted polymerisation. Hardware components like the electronic rack, the microscope and the xy-stage are indicated. The latter is a movable stage which serves for the coarse positioning of the substrate in the xy-plane. The positioning range of the stage comprises the hole substrate, but the accuracy is limited to approximately one micrometre. For a more precise positioning of the substrate, a piezo with a lateral driving range of 300 × 300 μm and an accuracy in the order of few nanometres is utilised.

Fig. 3.1 Direct laser writing system from Nanoscribe available at the Fraunhofer IPT

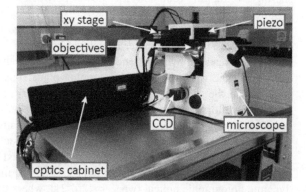

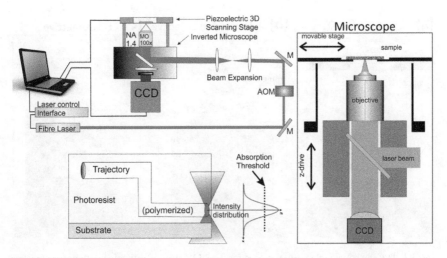

Fig. 3.2   Schematic sketch of the direct laser writing setup

The system features different writing modes, one standard method with an 100×
oil immersion objective and the "Dip In Laser Lithography"-objective which uses the
photoresist as an immersion fluid and thus allows to build up 3D-structures without
the laser light going through the already written structures. Therefore, structures of
over 1 mm in height may be written. The laser lithography system from Nanoscribe
(*Photonic Professional*) uses TPP in photoresists for the fabrication of true 3D nano-
and micro-structures. Where, e.g., conventional or electron beam lithography can
only realise 2-dimensional structures, direct laser writing additionally enables a fully
independent structuring in the third dimension and is thus also called "true" 3D
lithography. With this method, many applications like micro-optics, life sciences
and photonics can be addressed [23].

In Fig. 3.2 a simplified sketch of the setup is provided. A laser beam with a center
wavelength of 780 nm is focused on the sample. The laser used has a pulse length
of 150 fs, a repetition frequency of 100 MHz, and operates at a centre wavelength of
780 nm [24, 25]. Its intensity is controlled by an acousto-optic modulator (AOM).
A CCD camera enables a visual feedback according to the writing process. The
objective that focuses the laser beam can be moved in the z-direction in order to
adjust the position of the focal region.

The overview sketch in Fig. 3.2 is concretised in Fig. 3.3 with the illustration of
three different writing modes. These modes require three different objectives. An air
objective with a comparable large working distance of 1.7 mm is shown in (a). It can
be used for writing larger structures as the voxel size can exceed several micrometres.
In the case of a liquid resist, the latter is positioned on the substrate which then needs
to be transparent for exciting light. In (b) the setup using an oil immersion objective is
sketched. The resist is positioned on a glass substrate. The working distance is much
closer and immersion oil with a refractive index matched to the lens as well as to the
substrate is used as depicted in (b). A typical immersion oil has a refractive index

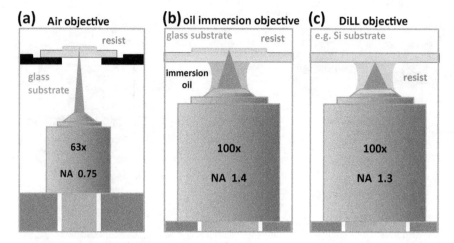

Fig. 3.3 Different Available writing modes for the DLW system. **a** Writing with an air objective, **b** using an oil immersion objective and **c** Dip-in Laser Lithography (DiLL), for which the photoresist simultaneously acts as an immersion oil

of around 1.5 and its positioning between a lens and a substrate of rather the same index increases the numerical aperture NA. The latter is given by $NA = n \cdot \sin(\alpha)$ where n is the refractive index and α is the angle spanned by the objective lens seen from the focal substrate. Hence, the NA of an air objective cannot exceed unity in contrast to an oil immersion objective. An increase of the NA positively affects the achievable resolution. A high resolution corresponds to a small resolving power $\tilde{\delta}$, which is given by

$$\tilde{\delta} \approx \frac{\lambda}{2 \cdot NA} \tag{3.1}$$

With the writing mode presented in (b) a lateral resolution of 500 nm and a lateral feature size of 200 nm can be achieved [23]. The limitations of this mode are met when the use of opaque substrates or structure heights exceeding the working distance are desired. Both challenges can be met by Dip-in Laser Lithography (DiLL), which is depicted in Fig. 3.3c. Here, the photoresist simultaneously acts as an immersion oil. Structures are therefore written top down which circumvents the limitation of the working distance. And as the light needs not to pass the substrate, the choice of the substrate material is less restricted.

3.5   Multifunctional Combined Microscopy Systems

Complete systems available for all three applications - HOTs, DHM-imaging or direct laser writing - are usually relatively compact devices which are solely restricted to one of these functions. If a system with increased functionality is needed in research, an individual unit of available upgrade modules has to be developed and adapted to a specific microscope as this has been performed in the past for a DHM-HOT setup [26]. Currently, to our knowledge no system is existing that offers additional holographic optical tweezers, simultaneous digital holographic imaging and multi-focal direct laser writing additionally to the possibility of bright field and fluorescence laser scanning microscopy in one setup. A system with such a range of functions would provide unprecedented possibilities to biomedical applications as well as photonics based on nano antennas. For example the positioning and manipulation of biological cells induced by optical tweezers combined with a digital holographic recording with simultaneous fluorescence imaging could provide a 2 1/2-dimensional multi-modal and thus high content image of the biological cell.

3.6   Nano Coordinate Measuring Systems

The increasing growth of the market for micro and nano system technologies is especially enhanced by expanded needs for components functionality for the consumer oriented market. For this market the cutting-edge technology originates from the optic and semiconductor industry, which have the highest demand in new definitions for quality control on micro systems with micro and nano metrology.

Metrological support for the measurement of various dimensional parameters or surface characterisations of the same micro-nano system is a necessity in research as well as for the development process.

In conventional metrology for objects beyond the size of typically several millimetres, a coordinate measuring machine (CMM) is utilised, featuring a tactile touch-probe and a precision positioning system with three orthogonal axes, often enhanced by rotational axes. Numerically controlled CMMs are very versatile and precise instruments, typically offering a precision in the micrometre range. However, when it comes to measuring microoptical or micromechanical components and workpieces, conventional CMMs often fail because of the following drawbacks:

- Conventional touch probes operate with contact forces in the order of at least mN. While this may be negligible for microscopic parts, a micro structured specimen may be deformed considerably.
- With a touch probe diameter of at least 1 mm tiny geometrical features of the workpiece cannot be resolved. Small gaps or boreholes are not accessible.
- Several significant measurement tasks cannot be solved in a tactile way. An example is the check of photo masks for lithographic processes that are plane chrome-on-glass structures.

Today, specially developed touch-probes for micro-metrology have diameters of 100–20 μm and some manufacturers offer new designed CMMs for micro-metrology [27–29]. But even these follow the basic construction principles with Cartesian linear axes. In contrast to this, atomic-force microscopes (AFMs) incorporate a tubular piezoelectric scanner, moving a small silicon tip with a radius of curvature in the

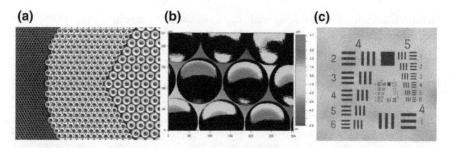

Fig. 3.4 Measurements performed with a CMM (NMM-1): **a** Microlens array with a pitch of 114.3, 255.6 and 400 μm (magnification: 50×), **b** white light interferometer measurement of the microlenses depicted in **a** with a radius of curvature of approximately 350 μm, **c** microscopic image of a 1951 USAF high resolution test chart (positive, chrome on glass)

order of nm in close proximity to the surface under test. With the probe tip of an AFM a nearly atomic resolution is attainable. Surfaces are scanned in a raster by a probe and topographies are sequentially built up. Commercially available instruments are limited to a measurement volume of less than $100 \times 100 \times 10 \, \mu m^3$ the natural sciences and widely used in semiconductor research, production as well as in the natural sciences.

The broad range of measuring systems utilised for metrological support for the micro and nano technology lead also to the problems of traceability and comparability of the measuring results. A combination of different measuring methods in one high precision multi-sensor device is an established technology for the acceleration of development processes and cost reductions because of the replacement of many different devices by only one multifunctional tool.

The most significant drawback of the commercially available multi-sensor-based systems is a restriction of the metrological quality of the whole device by sensor performance. Even if the main system, an AFM-sensor for instance, provides a suitable signal-to-noise performance, the AFM-sensor confines the measuring range and thus, the metrological quality of the result.

Optical techniques have limitations measuring steep surface slopes (Fig. 3.4a, b), specularly reflecting or transparent or black materials. For ceramics and plastics light is not reflected from the surface, but remitted from a volume below the surface, thus, causing erroneous optical measures.

To demonstrate the outstanding metrological capabilities of the instrument for different optics related specimens, several measurements were carried out to test the metrological characteristics of the system.

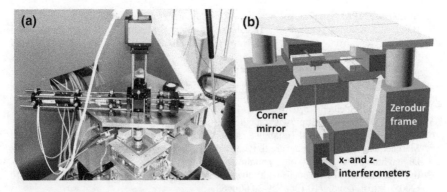

Fig. 3.5 **a** Setup of the NMM-1 with a mounted developed prototype sensor based on a low coherent interferometer in the centre of the stage, **b** 3D CAD-illustration of the metrological frame and the attached x, y, z-interferometers pointing into a virtual common centre surrounded by a corner mirror

3.6.1 Properties of the NMM-1 System

For the experimental investigations, especially in Sects. 4.3 and 4.7, a nanopositioning and measuring machine (NMM) is applied (NMM-1, SIOS, Fig. 3.5) which has been developed at the Institute of Measurement and Sensor Technology at the Illmenau University of Technology. It offers a positioning range of $25 \times 25 \times 5 \, mm^3$ with a resolution of 0.1 nm [27]. To use this highly precise long range NMM-1 as an instrument to measure micro- and nanostructures on small objects, different tactile and non-tactile optical sensors can be integrated into the NMM-1 [28] like a laser focus sensor, an AFM, a white light interferometer as well as a 3D micro probe. The position of the moving stage is measured and controlled by three fibre-coupled He-Ne-laser interferometers and two angular sensors. The stage itself consists of an extremely precise manufactured corner mirror on which the object to be moved is placed. It is positioned by electrodynamical drives, providing a continuously variable motion.

This device represents an Abbe offset-free design of three miniature interferometers, and applying a new concept for compensating systematic errors that result from mechanical guide systems, providing outstanding uncertainties of measurement. The high precision interferometer controlled stage defines the metrological performance of the whole measuring system and guarantees the traceability and repeatability of measuring results.

References

1. Elliot Scientific: E3300 Series Single Spot Tweezers Systems. http://www.elliotscientific.com/116-0-482-140/E3300-Series-Single-Spot-Tweezers-Systems/

2. Instruments, JPK: NanoTracker. http://www.jpk.com/nanotracker-overview.388.en.html
3. Bass, M., America, Optical S.: Handbook of Optics. McGraw-Hill Professional Publishing, New York (1994). http://books.google.de/books?id=B8UQSgAACAAJ, ISBN 9780070479746
4. Vogel, M.: Entwicklung und Aufbau einer modularen Konfokal-Multiphotonen-Laserscanning-Messapparatur (CMLTT) für Second Harmonic Generation-, Total Internal Reflectance und Laser Tweezers-Anwendungen an myofibrillären Präparaten. Dissertation, Ruprecht-Karls-Universität, Heidelberg 15.12.2004
5. Aresis: TWEEZ 200si. http://www.aresis.com/
6. Arryx: BioRyx 200 Optical Trapping System. http://www.arryx.com/product.html
7. Marquet, P., Rappaz, B., Magistretti, P.J., Cuche, E., Emery, Y., Colomb, T., Depeursinge, C.: Digital holographic microscopy: a noninvasive contrast imaging technique allowing quantitative visualization of living cells with subwavelength axial accuracy. Opt. Lett. **30**(5), 468–470 (2005). https://doi.org/10.1364/OL.30.000468
8. Imaging, Phase H.: HoloMonitor. http://www.phiab.se/
9. Tec, L.: DHM, The Digital Holographic Microscopes. http://www.lynceetec.com/content/view/474/184/
10. Restrepo, J., Garcia-Succerquia, J.: Automatic three-dimensional tracking of particles with high-numerical-aperture digital lensless holographic microscopy. Opt. Lett. **37**(4), 752–754 (2012)
11. Zhang, T., Yamaguchi, I.: Three-dimensional microscopy with phase-shifting digital holography. Opt. Lett. **23**, 1221–1223 (1998)
12. Cuche, E., Depeursinge, C.: Digital holography for quantitative phase-contrast imaging. Opt. Lett. **24**(5), 291–293 (1999)
13. Ashkin, A.: History of optical trapping and manipulation of small neutral particles, atoms, and molecules. Springer Series in Chemical Physics, vol. 67, pp. 1–31. Springer, Berlin (2001). https://doi.org/10.1007/978-3-642-56544-1_1, ISBN 978–3–642–62702–6
14. Svoboda, K., Block, S.M.: Biological applications of optical forces. Annu. Rev. Biophys. Biomol. Struct. **23**(1), 247–285 (1994). https://doi.org/10.1146/annurev.bb.23.060194.001335, PMID: 7919782
15. Fällman, E., Axner, O.: Design for fully steerable dual-trap optical tweezers. Appl. Opt. **36**(10), 2107–2113 (1997). https://doi.org/10.1364/AO.36.002107
16. Sasaki, K., Koshioka, M., Misawa, H., Kitamura, N., Masuhara, H.: Pattern formation and flow control of fine particles by laser-scanning micromanipulation. Opt. Lett. **16**(19), 1463–1465 (1991). https://doi.org/10.1364/OL.16.001463
17. Brouhard, G.J., Schek, H.T., Hunt, A.J.: Advanced optical tweezers for the study of cellular and molecular biomechanics. IEEE Trans. Biomed. Eng. **50**(1), 121–125 (2003). https://doi.org/10.1109/TBME.2002.805463
18. Dufresne, E.R., Spalding, G.C., Matthew, T., Sheets, A., Grier, D.G.: Computer-generated holographic optical tweezer arrays. Rev. Sci. Instr. **72**(3), 1810–1816 (2001). https://doi.org/10.1063/1.1344176
19. Dufresne, E.R., Grier, D.G.: Optical tweezer arrays and optical substrates created with diffractive optics. Rev. Sci. Instr. **69**(5), 1974–1977 (1998). https://doi.org/10.1063/1.1148883
20. Reicherter, M., Haist, T., Wagemann, E.U., Tiziani, H.J.: Optical particle trapping with computer-generated holograms written on a liquid-crystal display. Opt. Lett. **24**(9), 608–610 (1999). https://doi.org/10.1364/OL.24.000608
21. Curtis, J.E., Koss, B.A., Grier, D.G.: Dynamic holographic optical tweezers. Opt. Commun. **207**(1–6), 169–175 (2002). https://doi.org/10.1016/S0030-4018(02)01524-9, ISSN 0030–4018
22. Beck, R.J., Parry, J.P., MacPherson, W.N., Waddie, A., Weston, N.J., Shephard, J.D., Hand, D.P.: Application of cooled spatial light modulator for high power nanosecond laser micromachining. Opt. Express **18**(16), 17059–17065 (2010). https://doi.org/10.1364/OE.18.017059
23. Schaeffer, S.: Characterization of two-photon induced cross-linking of proteins. Master thesis, RWTH Aachen (2013)

24. Gansel, J.K.: Helical optical metamaterials. Dissertation, Karlsruher Institut for Technologie (2012)
25. Nanoscribe GmbH (2013). http://www.nanoscribe.de/de/produkte/photonic-professional
26. Barroso Peña, Á., Kemper, B., Woerdemann, M., Vollmer, A., Ketelhut, S., Bally, G.v., Denz, C., Popp, J., Drexler, W., Tuchin, V.V., Matthews, D.L.: Optical tweezers induced photodamage in living cells quantified with digital holographic phase microscopy. In: SPIE Photonics Europe, SPIE, 2012 (SPIE Proceedings), pp. 84270A–84270A-7
27. Jaeger, G., Manske, E., Hausotte, T., Buechner, H.-J.: The Metrological basis and operation of nanopositioning and nanomeasuring machine NMM-1. J. Vac. Sci. Technol. B (2009)
28. Manske, E., Hausotte, T., Mastylo, R., Machleidt, T., Franke, K.-H., Jäger, G.: New applications of the nanopositioning and nanomeasuring machine by using advanced tactile and non-tactile probes. Meas. Sci. Technol. **18**(2), 520 (2007). http://stacks.iop.org/0957-0233/18/i=2/a=S27
29. Dai, G., Koenders, L., Pohlenz, F., Dziomba, T., Danzebrink, H.-U.: Accurate and traceable calibration of one-dimensional gratings. Meas. Sci. Technol. **16**(6), 1241–1249 (2005). https://doi.org/10.1088/0957-0233/16/6/001, ISSN 0957–0233

Chapter 4
Experimental Methods and Investigations

Real-time high-throughput identification, screening, characterisation, and process-ing of reflective and transparent phase objects like micro and nanostructures as well as biological specimen is of significant interest to a variety of applications from cell biology, medicine and even to lithography. In this chapter, new as well as conventional possibilities that arise from digital holographic microscopy (DHM) and holographic optical tweezers (HOTs) are demonstrated and discussed on experimental setups and corresponding findings. This comprises subsequent digital holographic focus-ing, DHM with partially coherent light sources such as superluminescence diodes (SLDs) and LEDs, the tayloring of the coherence length of these broadband light sources as well as the simulation and the design of HOT setups. Furthermore, the characterisation of spatial light modulators and of an overall system is treated includ-ing the determination of the optical force and trapping stability. Exemplarily, also a DHM-HOT-system that is integrated in a nano-positioning system is demonstrated. In a further part of this chapter, the realisation and illustration of special beam con-figurations like non-diffracting light beams, that also allow exerting optical torque and new methods for assembling on the (sub)micrometre scale, is performed with focus on Bessel, Mathieu, Laguerre and Airy beams and their respective application for a DHM-HOT. An outlook to harness this setup for lithography, especially direct laser writing (DLW), is given in a last part of this chapter.

4.1 Objectives and Motivation

Optical microscopy is limited by the small depths of focus due to the high numerical apertures of the microscope lenses and the high magnification ratios. The extension of the depth of focus is thus an important goal in optical microscopy. In this way, it has been demonstrated that an annular filtering process significantly increases the depth of focus [1]. A wave front coding method has also been proposed in which a nonlinear phase plate is introduced in the optical system [2]. Another approach is based on a digital holography method where the hologram is recorded with a CCD

© Springer Nature Switzerland AG 2018
S. Stuerwald, *Digital Holographic Methods*, Springer Series
in Optical Sciences 221, https://doi.org/10.1007/978-3-030-00169-8_4

camera and the reconstruction is performed by a computer [3]. The holographic information involves both optical phase and intensity of the recorded optical signal. Therefore, the complex amplitude can be computed to provide an efficient tool to refocus, slice-by-slice, the depth images of a thick sample by implementing the optical beam propagation of complex amplitude with discrete implementations of the Kirchhoff–Fresnel (KF) propagation equation (2.45) [4, 5]. In addition, the optical phase is the significant information to quantitatively measure the optical thicknesses of the sample which are not available from the measurements with classical optical methods [6, 7]. Digital holographic microscopy (DHM) has been used in several applications of interest such as refractometry [8], observation of biological samples [9, 10], living cell culture analysis [11, 12], and accurate measurements inside of cells such as refractive indexes and even 3D tomography [13, 14]. As digital holographic methods allow to determine the complex amplitude signal, it is very flexible to implement powerful processing of the holographic information or of the processed images. As a non exhaustive enumeration of examples, in the following, methods are named for controlling the image size as a function of distance and wavelength [15], to correct phase aberration [16, 17], to perform 3D pattern recognition [18, 19], to process the border artefacts [6, 20], to emulate classical phase-contrast imaging [12, 21], to implement autofocus algorithms [22, 23], to perform object segmentation [24], and to perform focusing on tilted planes [25]. The optical scanning holography approach has been introduced with applications, for example, in remote sensing [26]. Holography is very often considered as a 3D imaging technique. However, the 3D information delivered by digital holography is actually not complete. This is the reason why optical tomographic systems based on digital holography were proposed [27, 28]. It also has to be mentioned that other 3D imaging can be achieved by the integral imaging method [29]. Therefore, digital holography provides not entire, even though to some extent, 3D information on the sample. Up to the present, a variety of holographic systems for microscopic applications have been developed for optical testing and quality control of reflective and (partially) transparent samples [11, 30–32]. Combined with microscopy, digital holography permits a fast, non-destructive, full field, high resolution quantitative phase contrast microscopy with an axial resolution <5 nm, which is particular suitable for high resolution topography analysis of micro and nano structured surfaces as well as for marker-free imaging of biological specimens [31, 32].

For optimised imaging conditions during long-term investigations, e.g. on reflective technical objects with dynamics along the optical axis - caused by the setup or thermal drift in machines - or on biological specimen, permanent focus tracking is required [33, 34] which is treated in a first subsection of this chapter. In the following sections characterisations of advanced DHM methods mainly based on low coherent light sources are presented that offer reduced noise but are challenging with regard to optimised utilisation. This also comprises investigations on adapted reconstruction methods and the demonstration of an enhanced imaging system for sub-micrometre topographies of technical and biologic specimen. In further works a DHM-system is extended by a HOT-system for simultaneous application of both digital holographic methods. For the handling of micro-particles and the efficient trapping of biologic cells, different tailored beams including non-diffracting beams are demonstrated.

4.2 Subsequent Digital Holographic Focussing

4.2.1 Autofocus Strategies and Application to Phase Distributions

The numerical analysis of digital holograms permits subsequent focus adjustment of a single captured hologram by the numerical propagation of the complex object wave based on the Fresnel–Kirchhoff diffraction theory [35]. Especially when utilizing LEDs as low coherent light sources for application in DHM [36, 37], a subsequent focusing is of special interest as coherence occurs only in a layer of less than $16\,\mu$m optical path length difference (OPD). Furthermore, the determined propagation distance to the sharply reconstructed image plane gives information about the axial object position and can be applied for monitoring of axial movements. The determination of the focused image plane is performed by the quantification of the image sharpness for every propagation distance and its maximisation. Considering the task of imaging regions of specimens with different objects at various axial heights, the focus plane position has to be determined for each region of interest. One numerical solution for decreasing the calculation time is to adjust the focus at a selected grid or set of regions and interpolate the focus for the rest of the image field. Problems may arise in case of an equal grid of points and partly periodic structures especially in case of technical specimens. To obtain a random set of points with low discrepancy here, the approach of a Halton set point is discussed in Sect. 4.2.2.

For a possible automation, robust image sharpness quantification algorithms are required. Therefore, four numerical methods for the determination of the optimal focus position in the numerical reconstruction and propagation of the complex object waves of phase objects are characterised, compared and adapted to the needs of digital holographic microscopy. Results from investigations of an engineered surface and human cells illustrate the applicability of Fourier-weighting- and gradient-operator-based methods for robust and reliable automated subsequent numerical digital holographic focusing. The application of the Forbes polynomial basis [38] is of particular interest to characterise and eliminate the aberration caused by the optical system in a more intuitively way and with less parameters in comparison to standard characterisations of optical aberration like the Zernike basis.

Autofocus Algorithms

Conventional microscopes offer a limited depth of focus which prohibits the observation of a complete image of a three-dimensional (3D) object in a single view. Auto-focusing algorithms based on image processing is a long-standing topic in literature. The focus-evaluation-function is a mathematical description used to measure whether the image is focused or not. Autofocusing is an essential feature for automated biological or biomedical applications like in autonomous micro-robotic cell manipulation as well as in high-throughput imaging in pharmaceutical research. Especially in case of imaging large objects, exact and reliable autofocusing techniques are fundamental for micro-electro-mechanical systems (MEMS).

Auto-focusing based on a pixel process method is a more flexible and preferred technology [39, 40]. Investigations have also led to the development of extended depth of focus algorithms for serial optical slices of microscopic 3D objects in recent years [41]. It takes advantage of the fact that an object appears sharpest when it is well focused. Otherwise, the image is blurry in out of focus position. A focus function is a quantitative description of the image sharpness in mathematics [42]. The maximum value of evaluation function indicates the focus location. If an evaluation function is applied to the focusing criterion, the curve of image sharpness evaluation function will have single peak. Some sharpness functions, such as image power functions, frequency domain functions, grey level variance measures, squared gradient function, gradient functions and information content functions, have been studied extensively in literature [39, 42, 43]. Till now, a wide variety of further focus algorithms have been investigated and characterised [44, 45]. In contrast to a mechanical variation of the distance of the microscope objective and the specimen, in digital holography the amplitude and phase distribution for different focus positions of the specimen can be calculated by altering the propagation distance d (2.45). Thus, the behaviour of the focus value can be determined numerically and subsequently without any need for additional mechanical devices. This concept is illustrated with an example in Fig. 4.1.

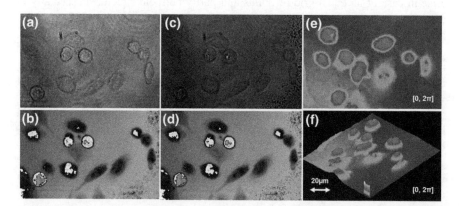

Fig. 4.1 One interferogram obtained from fixed biological cells is evaluated to obtain the amplitude- (**a**, **b**) and the phase distribution mod 2π (**c**, **d**). By means of digital propagation, structures in different object planes can be focused (**b**, **d**). After removal of the 2π ambiguity in the phase map (**e**) the unwrapped phase distributions is obtained, that corresponds - for the assumption of a constant integral refractive index - to the cells thickness. For adherent grown cells the false colour coded pseudo 3D representation represents the approximated cell shape (**f**)

Transparent objects (phase objects) are not visible in standard bright field optical microscopes. Often solely the edges of the object can be recognised. They can be visualised using the well-known Zernike's phase contrast microscopy, where the image contrast is proportional to the phase difference introduced by the object. If a phase object has uniform refractive index, phase-contrast microscopy then yields a measure of its thickness profile, making possible a quantitative reconstruction of the object from its image. In brightfield microscopy, pure phase objects like transparent specimens such as biological cells and technical reflective specimens are often better detectable when the microscope is slightly defocused, resulting in a minimal focus value when the object is exactly in focus [46]. This behaviour has to be taken into account for unwrapped phase distributions in digital holography as non continuous phase representations modulo 2π would be misinterpreted as sharp structures. This has been analysed comprehensively in [27, 47, 48]. For the described experiments, an FT-based unwrapping algorithm [49] is usually utilised in case of applying a recon-struction algorithm with no simultaneous unwrapping. To select robust and time-effective autofocus algorithms for application in digital holographic microscopy, four approaches that calculate a scalar focus value for image sharpness quantification are utilised for the comparisons of the approach with Halton-sampling and numerical parametric lenses (NPL). Thereby, the methods investigated for image sharpness determination of digital holographically reconstructed amplitude distributions were selected to fulfil the requirements for application in phase contrast microscopy [33]. Threshold-based algorithms [48] have been found to be not applicable for autofo-cusing in digital holographic microscopy, as the peak position of the focus function (threshold video-signal pixel count) or its width (threshold video-signal content) depend on the threshold value and therefore prohibit robust focusing. Furthermore, entropy-based concepts applied to an object wave $O(x, y, z)$ to retrieve a focus value $f(z)$ were reported to be not suitable for microscopic applications and thus were not investigated for the following characterisation [33, 46].

(I) Weighted Spectral Analysis

The weighted spectral analysis method quantifies the sharpness of the edges, as a focused image contains more fine details than a smooth defocused image. This parameter ($f\,d_{PS}$) is accessible by the measurement of the high-spatial-frequency content in the image, a measurement performed by summing up the logarithmically weighted and bandpass filtered power spectra [46]:

$$f_{PS}(z) = \sum_{x',y'} \log\left[1 + |\text{FFT}\left[|O(x, y, z)|\right]|\right] \tag{4.1}$$

where x' and y' denote the position in the Fourier domain of the amplitude distribution $O(x, y, z)$. The lower boundary of the applied band pass filter is set to suppress the frequency parts in the origin of the spectrum. Thus, the constant intensity offset in the image data is eliminated, as it contains no information about the focus degree. To increase the computation efficiency, the upper cut-off frequency is chosen in such a way that high-frequency parts of the images are suppressed that do not contain object information due to the lateral resolution of the optical imaging system.

(II) Variance of Grey Values

This method is based on the statistical analysis of the grey value distribution. Sharp structures in a focused image result in a higher contrast than in a smooth defocused image. The image contrast is quantified statistically by the variance (fd_{Var}) of the histogram of the grey level amplitude distribution $O(x, y, z)$ with the dimensions n and m and a mean value [27]:

$$f_{Var}(z) = \text{Var}\left[|O(x, y, z)|\right] \tag{4.2}$$

(III) Gradient Operator

The quantification of the focus degree of an image is performed by cumulated edge detection and thus by taking into account the increasing difference of neighbouring pixel values with sharper edges in the image. This principle is available by the calculation of the total absolute first derivative of the image by an absolute gradient operator (fd_{Gra}) [27, 46, 48]. For the discrete amplitude distribution, the first derivative is approximated by the difference quotient:

$$f_{Gra}(z) = \sum_{x=0}^{n-1} \sum_{x=0}^{m-1} \sqrt{\left(\frac{\partial |O(x, y, z)|}{\partial x}\right)^2 + \left(\frac{\partial |O(x, y, z)|}{\partial y}\right)^2} \tag{4.3}$$

(IV) Cumulated Edge Detection by Laplace Filtering

Applying the second derivative of the grey value distribution $e(x, y)$ especially quantifies the high-frequency content of an image by edge detection. As only changes in the gradient are taken into account, a constant grey value gradient is neglected. For the implementation of this method, the Laplace operator (fd_{LP}) is applied for the evaluation of the reconstructed amplitude distributions [43, 50]:

$$f_{LP}(z) = \sum_{x=0}^{n-1} \sum_{x=0}^{m-1} \left(\nabla^2 |O(x, y, z)|\right)^2 \tag{4.4}$$

Strategy for Search of Focus Plane

The determination of extrema in focus value curves within a propagation interval, for instance $Az = [-10, 10]$ mm that depends on the depth of field, is usually relatively inefficient. In order to speed up the process of locating extrema without loosing accuracy here first, a modified climbing-search strategy according to [49] is implemented comprising gradient and iterative approaches with a refined determination of the focus values. Alternatively, a combination of bisectioning and bicubic interpolation [51] is utilised, which leads often to a shorter calculation time.

Numerical Phase Masks in Forbes Representation

The reconstructed object wave in the hologram plane $O(x, y, z_0)$ is often distorted caused by aberrations in the optical system. The concept of numerical parametric lenses (NPL) is a common feature in DHM to achieve optimised wave front reconstruction in digital holography [52]. Operations usually performed by optical components and described in ray geometrical optics, such as image shifting, magnification, and especially complete aberration compensation (phase aberrations and image distortion), can be mimicked by numerical computation of a NPL. Furthermore, automatic one-dimensional or two-dimensional fitting procedures allow adjustment of the NPL parameters as expressed in terms of standard or Zernike polynomial coefficients. These coefficients can provide a quantitative evaluation of the aberrations generated by the specimen. Therefore, (2.45) can be rewritten as follows:

$$O(X, Y, d) = \Gamma^I(X, Y) A \times \text{FFT}^{-1} \left\{ \text{FFT} \left[\Gamma^H(x_0, y_0) \, O(x_0, y_0, z_0) \right] \times \exp\left(i\pi d \left(\xi^2 + \eta^2\right)\right) \right\}$$

$$(4.5)$$

with the two NPL-arrays of unit-amplitude complex numbers Γ^I and Γ^H placed respectively in the image or hologram planes. Since a Zernike representation needs a relatively long computation time and is relatively complex in higher orders, we propose the application of the more intuitively Forbes polynomial basis [53]. These are generated by normalizing the transverse variable in the standardised representation of aspheres to $u = r/r_{max}$:

$$z = \frac{cr^2}{1 + \sqrt{1 - \epsilon c^2 r^2}} + u^4 \sum_{m=0}^{M} a_m Q_m^{con}(u^2) \, , \qquad (4.6)$$

where r denotes the radial distance in the plane perpendicular to the optical axis z (see also Appendix B.3.9). This orthogonal basis with the members Q has the advantage, that in case of fitting to given shapes, the optimal value of any coefficient is independent of M. Furthermore, the mean square value of the additive part in (4.6) is just a weighted sum of squares, which avoids cancellation problems and allows to interpret the coefficients more intuitively.

4.2.2 Halton Sampling

When considering the task of measuring the complex object wave, in particular the phase distribution, of a large xy-plane of a specimen by stitching the fields of view, the focus plane position needs to be determined for each subregion with a sufficient spatial sampling frequency. For a 10×10 field region, this implies e.g. repeating the autofocus procedure 100 times, which requires a large amount of time. Furthermore, in case of fluorescence imaging, the sample fluorochrome ability to emit fluorescence is negatively affected. To overcome these problems, a possible solution is to refocus at a few positions within the region of interest and estimate the rest by application of interpolation. While literature concerning focusing algorithms is rich, the way the points are chosen is not covered sufficiently yet to our knowledge for microscopy. Especially when measuring (partly) periodic structures, an equally spaced grid of points for subsequent numerical determination of the focus plane is often not advantageous in case the spatial frequency of the focus grid is approximately a multiple integer of the spatial frequency of the specimen.

The simplest way to sample the xy-focus positions is e.g. to select one or more positions for every field of view in each direction, thus resulting in a equally spaced grid-like sampling scheme. There are many other possible schemes, but optimised results with less time consumption are only achievable with an appropriate choice of the sampling strategy.

In general, it is assumed that by sampling with a high degree of uniformity, the interpolation accuracy can be improved for a majority of specimen. Thus, a discrepancy measure of a certain set as a measurement of the degree of uniformity of a candidate distribution of points is utilised for a characterisation. Since the distribution of cells and its sizes often show special distributions, a slightly non uniform sampling point distribution can have the potential to reduce the - application dependent - required effective total density of sampling points.

Till now, the concept of discrepancy in numerics has been designed and studied for various point distributions to achieve low discrepancy. Application of such sets cover many differing fields, such as cryptography or numerical integration. Here, a low discrepancy set called the Halton point-set is selected [54–57] since common randomly chosen numbers or positions can have a relatively high discrepancy. Discrepancy can be formally defined in many divers ways. Here, the most common variant is utilised, which is the discrepancy with regard to axis-parallel boxes. It is determined as the maximum absolute difference between the expected fraction of points falling in an axis-parallel box - a box with its edges parallel to one of the axes - and the fraction of the points it actually contains. A few examples in two dimensions may help to illustrate the definition:

In case there is a single sample at the corner of a rectangular region, the discrepancy of this point set would be 1: The maximised box one can build is a box containing the whole rectangle except the corner point and while this box is empty, one would have expected to find the totality of the points. If there is one sample at the centre of the region, the discrepancy already decreases to 0.5. Simplified expressed, at the limit, if the number of samples to cover the entire region with points is expanded, then the discrepancy of the set would be 0. This means, the better the points are laid out, the lower the discrepancy of the point set will be. Two dimensional grid-like sampling results in a discrepancy proportional to the square root of the number of samples. The Halton sampling set's discrepancy is upper-bounded by the algorithm according to the following relation:

$$L_{\text{Hal}} \leq \frac{ljn(N_s)}{N_s} \tag{4.7}$$

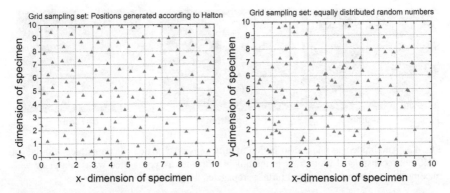

Fig. 4.2 Left: Axis parallel grid representing the field of view of 10×10 and the corresponding Halton point-set. Right: Statistically uniform distributed random numbers for xy-focus position determination with the property of higher discrepancy

Even for low number of samples N_S, one can see that the discrepancy of the Halton point-set is significantly lower than the grid point set. Motivated by these considerations, it is investigated if the utilisation of this sampling strategy would improve the result of the interpolation.

4.2.3 Experimental Investigations

The autofocus algorithms applied on the retrieved phase and amplitude data of the scanning process of the propagation are demonstrated on a coated silicon waver surface that is structured with geometric patterns of 50 nm height. Figure 4.3 shows the results of investigations of the nano-structured surface in reflection mode with a *Zeiss LD-Achroplan* ($20\times$, $NA = 0.40$, corrected) microscope objective, where the object is imaged slightly defocused onto the CCD. For purpose of comparison, a 1951 USAF test chart is shown with the correspondent focus position. In case of unfocused imaging the surface structure in the reconstructed amplitude distribution nearly disappears [Fig. 4.3a, c], whereas the unwrapped phase distribution in Fig. 4.3b contains more sharp structures and diffraction. In both symmetric defocused cases the structures in the amplitude distribution show diffraction fringes [Fig. 4.3a, c], and the phase distribution shows a decreased numerical sharpness value [Fig. 4.3a, c] in contrast to the amplitude behavior which usually has a maximum (see also Fig. 4.14). In Fig. 4.4 the corresponding normalised focus value functions for fd_{SP}, fd_{Var}, fd_{GRA}, and fd_{LP} are depicted. It can be seen, that except the variance is less suitable for the focus determination of phase distributions since the extremum is relatively unsharp. An example for defocused living cells is depicted in Fig. 4.5.

Application of Halton Sampling

The Halton point-set is investigated with simulations based on a hologram of three kinds of surfaces, two of them are artificially generated. Thereby, the focus plane

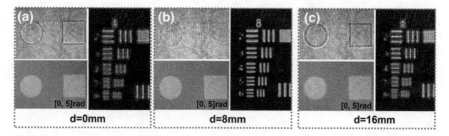

Fig. 4.3 Amplitude and phase distribution of structures on a coated silicon surface with height of 50 nm: reconstructed wave front in the hologram plane (**a**), propagated wave to a plane with maximum image sharpness (**b**) and further propagated image in (**c**)

Fig. 4.4 Focus values of unwrapped phase distributions of the specimen in Fig. 4.3: A clear dip in the center demonstrates the global minimum of the focus algorithms except the variance $f d_{Var}$

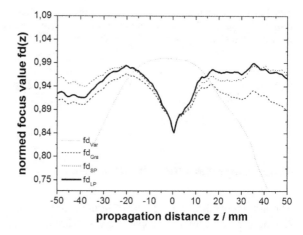

was chosen to follow a fractal surface [57]. The focus data is generated by repeating an autofocusing algorithm for each region of interest within a specific xy-region intended to be measured. The real specimens are living tumor cells and depicted in Fig. 4.6.

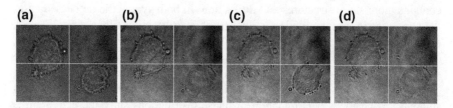

Fig. 4.5 Amplitude distributions of tumour cells with an overlaid coarse grid for illustrating the regions of interest applied for the Halton point-set. **a** shows an unfocused amplitude distribution, in **b** and **c** only one of the cells is imaged sharply by propagation of the complex wave. In **d**, both cells are propagated with individually optimised propagation distance d to maximize the image sharpness by utilisation of Halton sampling and an interpolation of the propagation distance

In order to quantify the impact of the sampling strategy, the following procedure for generating the sample points and measuring the error is performed: First, a few sample distributions are generated in order to be compatible among the sets. In fact, while Halton point-sets can be created of any size, the number of points in a grid-like point set is not random: In case of a square region, one can only generate point-sets which have a number of points which is a perfect square. If incomplete grids are created, the discrepancy is automatically increased by an amount that is directly proportional to the sum of points needed to fill in the nearest complete grid. One example of two comparable sets is illustrated in Fig. 4.2. In a second step, the area is sampled at the positions specified by the point set. After the sample is recorded, Gaussian noise is added with a mean equal to zero and a variance large enough to attain a specified signal-to-noise-ratio (SNR). This has to be tested over a range of several different SNR levels (here approx. 6–8). Thirdly, the samples are used to create a new surface by means of interpolation. The positions where the interpolation is performed are those where the original surface was specified, thus allowing a calculation of a meaningful difference. Finally, the magnitude of the difference between the interpolated and original surface is defined as the interpolation error. Therefore, two cumulative figures are calculated to quantify this error: the cross-correlation coefficient and the sum of the squared differences. To test the statistical significance between using different sampling schemes, a non-parametric test for difference of means is applied by considering the sum of the squared differences (normalised mean square error) as a mean and comparing the two means to see if the difference has been significant.

In Fig. 4.6 a synopsis of the performed investigation is given by analysing 10 different images and comparing the focus value in case of a Halton set and a uniform sampling for different sampling densities on the images. It can seen, that for a certain sampling density (or division factor for the ROI of an image) which is in the area of $\approx$[9, 21] the Halton sampling delivers slightly better results than a simple uniform sampling distribution.

In the framework of the numerical investigations it has been revealed, that for digital holographic microscopy with regional optimised propagation distances in $57\% \pm 3\%$ of all tested combinations of noise, surface or background types and sampling density, Halton sampling achieved the lowest error after interpolation in this study. In case of the biological cells depicted in Fig. 4.6, an improvement in average of 45% of the cases has been attained which, however, depends on the type of cell which has to be imaged.

Conclusion It is demonstrated that common autofocus algorithms are not only applicable to amplitude representations, but also to phase distributions. In case of phase distributions, the focus curve shows a minimum amplitude image sharpness due to the effect of diffraction fringes in the reconstruction and unwrapping process. The application of Halton sampling in digital holographic microscopy allows - with regard to equally distributed random numbers - a more efficient automated and subsequent region selective refocusing process especially in case of periodic specimens. The application of these focus quantification methods is also demonstrated in Fig. 4.14.

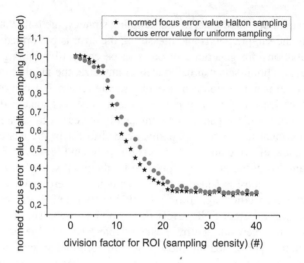

Fig. 4.6 Amplitude distributions of tumour cells with an overlaid coarse grid for illustrating the regions of interest applied for the Halton point-set. **a** shows an unfocused amplitude distribution, in **b** and **c** only one of the cells is imaged sharply by propagation of the complex wave. In **d**, both cells are propagated with individually optimised propagation distance d to maximize the image sharpness by utilisation of Halton sampling and an interpolation of the propagation distance

Furthermore, the aberration compensation of the optical imaging system by numerical parametric lenses in Forbes representation is simpler and less time consuming in calculation.

4.3 Digital Holographic Microscopy with Partially Coherent Light Sources

In this section the utilisation of partially coherent illuminations for digital holographic microscopes (DHMs) in transmission geometry is investigated. The application of different low coherent light sources such as superluminescent diodes (SLDs), LEDs and supercontinuum light sources is characterised and optimised. Therefore, new approaches for reconstruction of the complex wave front are developed theoretically and demonstrated experimentally. Depending on the requirements given by the application, the light fields can also be generated by a spatially filtered LED or from a light beam transmitted through an adapted frequency (wavelength) filter. A major advantage is the significant reduction of the speckle noise which opens up an improved image quality and the proper evaluation of phase contrast modes such as differential interference contrast (DIC) and digital holographic imaging modes. For biomedical applications, in this work a laser scanning and digital holographic microscope is coupled with fluorescence sources to achieve multimodal diagnostics. Several implementations of biomedical applications where digital holography with low coherent light is a significant improvement are described and demonstrated. With a fast DHM permitting the analysis of dynamical phenomena, several applications in fluid physics and biomedical applications can also be provided. Typically, digital holography is performed with coherent laser beams that suffer from their sensitivity towards any disturbances in the optical paths like straylight leading to lower imaging results mainly caused by the coherent artifact noise. Therefore, different DHM-setups are implemented for this work with partially spatial coherent illumination in a Mach–Zehnder and a Linnik configuration in order to reduce this noise [12, 58–60]. The first DHM is using a standard single emitter LED in an epoxy dome lens that is spatially filtered to achieve the sufficient spatial coherence. Since the measured spectral bandwidth is ≈20 nm, the source is also of partially temporal coherence. This characteristic is advantageous towards the noise reduction, but it requires to precisely align and adjust the reference and object beams [61]. Thereby, the calculation of the optical phase is highly affected by the implementation of a proper or specially adapted phase shifting method (see Sect. 2.2). For applications where the OPD or the specimen's slope is varying too rapidly to implement a temporal phase shifting, a DHM where the complex amplitude is computed with the Fourier transform method [62, 63] - also known as a variant of spatial phase shifting - is implemented. In this case, there is an average angle between the reference and the object beam when superposing both waves that requires coherence leading to a sufficient amount of interference fringes for the algorithms to work. With respect to that constraint, it is also possible to realise a partially spatial (but also temporal) coherent source from a coherent laser beam focused on a rotating ground glass. The advantages of using the partial spatial coherence in a DHM setup are discussed in Sect. 4.3.3.

4.3.1   Optical Setups and Digital Holographic Reconstruction

Digital Holographic Microscopy Based on LEDs and SLDs

The DHM-system that benefits from a partially spatial coherent illumination created e.g. by a superluminescence diode (SLD) is described in Fig. 4.7. For exploiting a maximum of the LED light emission, an LED beam is injected in a liquid optical fibre in order to homogenize the beam distribution in the front focal plane of a lens which is imaged in the plane of the sample. The sample may also be placed - in contrast to Fig. 4.7 - inside a Mach–Zehnder interferometer in order to record the interference patterns between the sample beam and the reference beam. In addition to the partially spatial coherent nature of the illumination, the LED or SLD provides also a low temporal coherence. For samples embedded in scattering media like collagenous gels, this reduces further the noisy coherent contribution by selecting the ballistic photons for the interferometric information. The mirror is placed on a piezoelectric transducer to implement temporal phase stepping method.

The system provides the optical phase and the intensity images on one plane of a sample, with a typical acquisition time of $1/4 s$ due to a low intensity and a

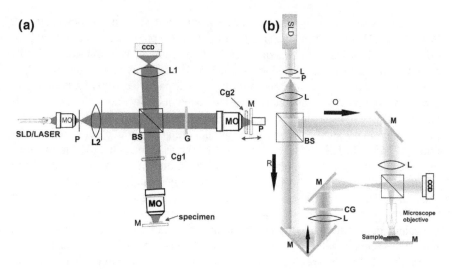

Fig. 4.7 Experimental setups for low coherent light sources: Left: Linnik-type interferometer with a SLD (LED) as light source, a spatial filter (P, L), a beam splitter cube BS, mirrors M, piezo actuator PZ, two identical microscope objectives MO $(20\times, NA = 0.2)$, a CCD-camera and optional dispersion compensation glasses Cg1 and Cg2. Right: Mach–Zehnder interferometer with a variable delay line by variation of the position of the mirror M and a compensation glass CG

required minimum integration time. The resulting complex optical field can then be utilised to refocus the optical intensity and phase fields on parallel planes without any mechanical scanning or loss of time. As the required information for performing refocusing is recorded only in the recorded plane by the CCD, the point spread function of the optical system is constant over the experimental volume, provided that the digital holographic refocusing does not introduce a loss of information. The magnification corresponding to that of a standard optical microscope is about $100\times$, the lateral resolution computed according to the Rayleigh criteria is $1.3\,\mu$m, and the Z resolution, denoted here with $\bar{\delta}$, computed with the standard formula $\bar{\delta} = \lambda/NA^2$ is $5.4\,\mu$m, where λ is the average wavelength of $660\,$nm and NA the numerical aperture of 0.30. The depth of refocusing by means of digital holographic reconstruction is extended by a factor of about 100 [58]. The resolution of the optical thickness computed on the phase map is about $2\,$nm. This value is established by assessing the noise that occurs on the phase maps. As the full information on the optical field transmitted by the sample is recorded, it is possible to emulate standard optical microscopy modes such as the differential interference contrast (DIC). This latter is particularly useful for the observation of living cells by providing the scientists with an usual visualisation mode. It has to be considered that the DIC mode is successfully implemented thanks to the low noise level. Furthermore the digital holographic microscope is coupled with a fluorescence excitation source (eGFP) and, as shown in Sect. 4.3.3, it is possible for refractive samples to couple the fluorescence and the digital holographic signals to perform refocusing.

4.3.2 Coherent Noise Removal

After the benefits of low coherent light have been demonstrated theoretically in Sect. 2.4, some experimental demonstrations will be treated in the following.

A major issue of the full coherent illumination is the coherent speckle noise that arises on the unavoidable defects of the optical system, on the sample container that can be in some cases of poor optical quality, and on the sample itself when it is bulky and scatters the light. The speckle results from the coherent superposition of random contributions that originate from out of focus locations. A way to reduce the strength of the speckle noise is to reduce the spatial coherence of the beam illuminating the sample. The relation between the noise and the spatial and temporal coherence states, although not performed in the frame of digital holography, has already been shown many years ago [64].

Therefore, an example from a fluid physics experiment will be given here. The images in Fig. 4.8 are magnified parts of digital holograms that were recorded with a partially spatial illumination and a fully coherent illumination. The object is identical and consists of 5 μm, particles immersed in distilled water. In the partially spatial coherent case, one observes that the diffraction pattern of the unfocused particles is clearly distinguishable. As a result, the digital holographic reconstruction can be applied to refocus the particles in the field of view. On the contrary, with the full coherent illumination, the only visible particles are those that are focused and the background is a speckle field. In that case the digital holographic refocusing is very noisy. The applied microscope lenses are 20× microscope lenses (*Zeiss Achroplan*) with a numerical aperture of 0.3. The effective magnification is 15× to provide a 390 μm × 390 μm field of view on the CCD sensor. According to [12, 65], the maximum refocusing distance is approximately ±250 μm around the best focus plane in the coherent case without significant loss of resolution.

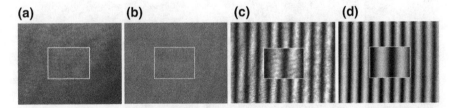

Fig. 4.8 Digitised interferograms in a 256-bit grey-scale format illuminated with coherent He-Ne laser light in **a**, **c** and low-coherent light of a red LED in **b**, **d**. The interferograms **a**, **b** are shifted in time and in **c**, **d** spatially shifted. The white outlined area represents respectively an enlarged section. Comparing the standard deviation in a plane area, the noise with a laser is approx. three times higher than with LEDs

4.3.3 Experimental Demonstrations and Applications

Biomedical Applications

The biomedical DHM-applications that are presented in the following have been performed with a LED illumination as described in Fig. 4.7.

Study of Cell Cultures

The DHM-system that has been optimised in this work is a powerful tool to study cell cultures and their evolution in time. Thanks to the partially coherent source, the holographic microscope provides a bright field image quality equivalent to a classical microscope, even with sample container of poor quality such as plastic dishes (Fig. 4.9). As the full information on the optical field transmitted by the sample is recorded, it is possible to emulate classical optical microscopy modes such as differential interference contrast (DIC) to improve the visibility of unstained living cells [12] (Fig. 4.9). The quantitative phase computation is used to analyse dynamically the cell morphology with nano-metric accuracies.

4.3.4 Adaption of Reconstruction Methods

Here, the effects on the reconstruction quality by considering the whole spectrum of the light source are investigated. Furthermore, the propagation of the complex

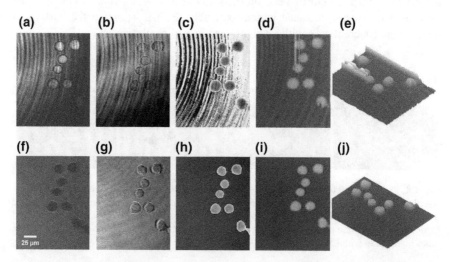

Fig. 4.9 Digital holographic investigation on pancreas tumour cells; illumination with laser light (upper row) and LED light (bottom row). **a, f** Holograms (exemplarily of one of three recorded temporal phase-shifted holograms); **b, g** amplitude distribution; **c, h** phase distribution modulo 2π; **d, i** unwrapped phase distribution; **e, j** grey-level-coded pseudo-three-dimensional representation of the phase maps in **d** and **i**

wave fronts, which are determined by digital holography, and used for re- and multi-focusing, is modified to light sources with a spectral width wider than that of typical laser light sources. Therefore, the propagation algorithm using the convolution method as a solution of the Fresnel–Kirchhoff diffraction integral according to the first Rayleigh–Sommerfeld approximation is extended by an additional integral to take into account the spectral width of low coherent light sources. Numerically, in the new approach the supplementary integral is realised by a discrete sum considering a finite set of wavelengths. Specifically, the results of the modified algorithm are compared with common algorithms with respect to resolution and image sharpness.

For optimised imaging conditions during automated measurements for quality control, on objects with dynamics along the optical axis or long-term investigations, e.g. on living biological specimen in climate chambers, permanent numerical focus correction is an advantageous feature of digital holography [33]. The numerical analysis of digital holograms permits subsequent focus adjustment of a single captured hologram by the numerical propagation of the complex object wave [66]. To reduce the signal noise caused by multiple reflections and stray light in the setup, short coherent light sources are utilised [67]. Till now, also the propagation of these complex wave fronts is performed by using the central wavelength of the broad spectrum of the light source. Here, a propagation considering the spectral distribution of the low coherent light source and therefore a superposition of complex waves propagated with different wavelengths is proposed. The new approach is finally applied for imaging a selection of test charts, but also microlens arrays.

Reconstruction with Consideration of Spectrum

When recording digital holograms with light sources like LEDs which have spectral ranges of several tens of nanometres (typically $\approx 0.20\,\mu m$), the complex wave is - in first order - reconstructed and propagated with the central or main wavelength. Thus, a superposition of complex waves propagated to the image plane with consideration of the normalised spectral distribution $P(\lambda)$ of the light source is proposed. The same concept is applicable for multi-wavelength digital holography if using a tunable laser light source for example, but not demonstrated here. For most light sources with a broad spectral range, a Gaussian spectral distribution can be assumed (see e.g. Fig. 4.10), which corresponds to a weighting function in the following calculations:

$$P(\lambda) = \frac{1}{\sqrt{2\pi}\sigma} \exp\left(-\frac{(\lambda - \lambda_c)^2}{2\sigma^2}\right) \tag{4.8}$$

If considering the spectral distribution, (2.45) has to be extended by an additional integral concerning the wavelength. For a numerical realisation, the integral is replaced and approximated by a discrete sum of N complex waves:

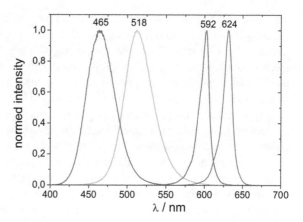

Fig. 4.10 Spectra of utilised 5 mm standard LEDs as light sources in the digital holographic setup

$$O'(x, y, d) = \int P(\lambda) \cdot e^{\left(i\frac{2\pi}{\lambda}\frac{x^2-y^2}{2d}\right)} \times \text{FT} \left[O(\vec{r_0})e^{\left(i\frac{2\pi}{\lambda}\frac{x_0^2-y_0^2}{2d}\right)} \right] * \text{sinc}\left(\frac{nx}{\lambda d}\right) \text{sinc}\left(\frac{my}{\lambda d}\right) d\lambda$$

$$\approx \sum_{j=1}^{N} P'(\lambda_j) \cdot e^{\left(i\frac{2\pi}{\lambda_j}\frac{x^2-y^2}{2d}\right)} \times FT \left[O(\vec{r_0})e^{\left(i\frac{2\pi}{\lambda_j}\frac{x_0^2-y_0^2}{2d}\right)} \right] * \text{sinc}\left(\frac{nx}{\lambda_j d}\right) \text{sinc}\left(\frac{my}{\lambda_j d}\right)$$

$$(4.9)$$

with the discrete normalised spectral distribution $\sum_{j=1}^{N} P'(\lambda j) = 1$.

The noise, but not the amount of noise in the hologram slightly differs with respect to the chosen wavelength, as the complex field of the diffracted wave depends on the wavelength based on (2.45). Here it is assumed that the noise is randomly changed in relation to the wavelength. Namely, the intensity of the reconstructed object $I'(x, y, z)$ is assumed to be given by:

$$I'_{\lambda_j}(x, y, z) = I_{\lambda_j}(x, y, z) + s_j(x, y, z) \qquad (4.10)$$

where $I_{\lambda_j}(x, y, z)$ and $s_j(x, y)$ denote the intensity of the reconstructed object without noise and the noise pattern reconstructed by the wavelength λ_j respectively. In case of summation of $I'_{\lambda_j}(x, y, z)$ over different wavelengths, we have

$$\sum_{j=1}^{N} I'_{\lambda_j}(x, y, z) = \sum_{j=1}^{N} P'(\lambda_j) \left[I_{\lambda_j}(x, y, z) + s_j(x, y, z) \right],$$

$$\text{with } \sum_{j=1}^{N} P'(\lambda_j)s_j(x, y, z) = S \qquad (4.11)$$

where the signal noise S is assumed to be constant. As stated above, this equation is based on the assumption of randomly changed noise if the propagation is performed with a different wavelengths. If the constant value S is very small in comparison to $NI'(x, y, z)$, the noise can be neglected in first order and the following relation is obtained:

$$I'(x, y, z) = I(x, y, z) \tag{4.12}$$

Therefore, in an analogous way, by superposing the multiple reconstructed complex waves with different wavelengths, the noise is smoothed by an averaging effect so that the image quality can be improved. The improvement of the reconstruction if using a complex wave addition is compared in the next subsection with the addition of the amplitude and phase alone.

If superposing complex waves which are reconstructed with different wavelengths, it has to be taken into consideration that the pixel sizes Δx and Δy of the reconstructed image vary according to the wavelengths and are given by

$$\Delta x = \frac{\lambda d}{N_x \Delta X}, \quad \Delta y = \frac{\lambda d}{N_Y \Delta Y} \tag{4.13}$$

respectively, where N_X and N_Y denote the number of pixels of the digital holograms, respectively, and ΔX and ΔY denote the pixel sizes of the imaging devices [68, 69]. The pixel size in the reconstructed image should be equal to superpose. For reasons of simplicity, the following considerations are solely performed for the x-dimension. To adapt and equalize Δx, one of the parameters d, N_X, or ΔX have to be modified. One reasonable approach is to alter N_X for a different wavelength by padding zeros [69] which needs neither experimental manipulation nor signal processing. For fast implementation of the propagation, the fast Fourier transform (FFT) algorithm is utilised for which the sampling number is usually equal to the Nth power of two.

Here, the pixel size (sampling interval) Δx of the digital hologram is changed by compensation (Fig. 4.11). Denoting λ_c a criterion wavelength and λ_i ($i = 1, \dots, N$) representing a wavelength applied for recording. With (4.13), the compensated sampling interval $\Delta x'$ yields:

$$\Delta x' = \frac{\lambda_i}{\lambda_c} \Delta x \tag{4.14}$$

Therefore, the pixel value (e.g. amplitude A and phase ϕ_O) of the hologram has to be interpolated between adjacent pixel values. Here, a linear interpolation is chosen to determine the value of the compensated pixels. It is assumed, that the change of the pixel value of the interferogram (complex digital hologram) is not sharp. Let for instance $A(k)$ and $A'(l)$ denote the amplitude of the digital hologram at the k-th pixel and the amplitude of the compensated digital hologram at the lth pixel respectively. The schematic diagram of compensation of a digital hologram for a different wavelength is shown in Fig. 4.12. The compensated pixel value $A'(l+1)$ is then given by:

$$A'(l+1) = \frac{A(k+1) - A(k)}{\Delta x} \left[(l+1)\Delta x' - k\Delta x \right] + A(k) \tag{4.15}$$

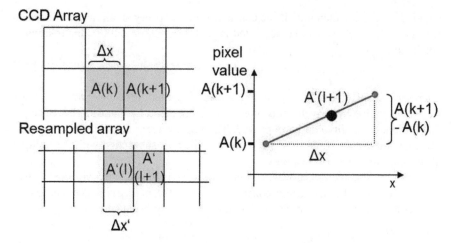

Fig. 4.11 Illustration of the resizing procedure of a propagated hologram to adopt the resolution for a superposition of multiple holograms propagated with different wavelengths

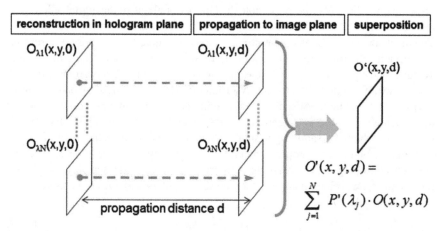

Fig. 4.12 Illustration of the discrete, wavelength selective reconstruction and propagation process with final superposition of waves resulting in $O'(x, y, d)$

Experimental Investigations

To confirm the proposed method, the experimental results are demonstrated utilizing the experimental setup depicted in Fig. 4.7. It shows a Linnic-type setup adopted for temporal phase shifting digital holography by integrating a piezo driven mirror. Here, a three step algorithm is applied with optimised parameters according to [17]. Standard 5 mm LEDs are used as light sources. The CCD camera has a resolution of 1024×768 pixels with a pixel pitch of $4.65\,\mu m$. For magnification, $20\times$ microscope objectives are used ($NA = 0.2$). The corresponding spectra with a Gaussian distribution are shown in Fig. 4.10. The widths (FWHM) of the spectra are in the range of 13–31 nm whose properties are investigated in [18]. For the investigations on the wavelength selective algorithm according to (4.9) a USAF 1951 resolution test chart is applied. The new propagation method has been used for improvement in imaging microlens arrays on silica substrate with a pitch of approximately 114 μm. In Fig. 4.13 the reconstructed USAF 1951 test chart is shown in the hologram plane. The superposition of $N = 20$ propagated holograms according to (4.9) is compared with a single hologram propagated with the central wavelength λ_c Additionally, instead of complex addition of the waves, an addition solely of the amplitudes of the propagated holograms has been performed for a comparison. The normalised noise σ_N represented by the standard deviation within the white rectangle of every prop-agated image in Fig. 4.13 is compared. The values are $\sigma_N = 1$ for (b), $\sigma_N = 0.62$ for (c) and $\sigma_N = 0.48$ for (d). It can be concluded, that the hologram propagated with a single wavelength in Fig. 4.13a has a higher amount of artefacts consisting of a parallel pattern in comparison to Fig. 4.13c, d. This is due to aliasing which is a significant error source if applying long propagation distances d. Reconstruction with consideration of the spectrum of the light source - as depicted in Fig. 4.13c, d - leads to less aliasing mainly due to the smoothing effect of the superposition of the different holograms. Here, the complex addition of the holograms leads to low-est noise and highest contrast in Fig. 4.13a in comparison to the other methods. It is observed that the applied USAF test chart is in focus at a second propagation distance which is not shown here. This is caused by a second image as a result of reflection in the Linnic type interferometer with the transparent USAF test chart. Because of this property, this test object is ideal for a comparison of algorithms in order to evaluate the performance of the algorithms in presence of additional patterns.

The corresponding values for the image sharpness as function of the propa-gation distance according to the focus value algorithms in Sect. 4.2.1 are shown in Fig. 4.14a–d. Every focus algorithm shows a maximum value at a distance of $z = 10$ cm. The second, but lower peak at the distance $z = 17$ cm corresponds to the second image due to the reflection at the mirror in the setup causing a second sharp image (see Fig. 4.13). The focus value in case of complex summation are in general higher than those in case of amplitude summation, which corresponds to the find-ings in Fig. 4.13. Propagation only with the central wavelength has the highest focus values. But with regard to Fig. 4.14b, this is explained with the effect of aliasing, which causes a sharp parallel line pattern. Furthermore, here maxima of the focus values are less sharp and thus sometimes harder to find. Therefore, the new method for propagation increases the robustness of the measurement process if an automated measurement by machines is performed on micro-optical components, where a sub-sequent focus or a correction of the image sharpness after the measurement is of particular interest. Additional, the determined propagation distance to the sharply reconstructed image plane gives information about the axial object position and can be applied for monitoring of axial movements, e.g. in a production machine. These advantages are useful for imaging microlens arrays with LED light, as depicted in Sect. 4.8.

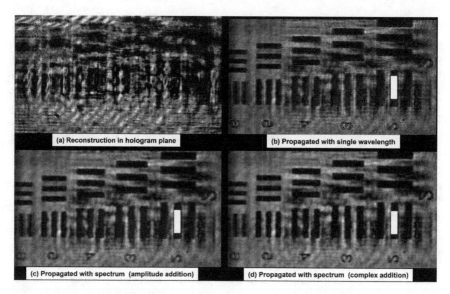

Fig. 4.13 Reconstructed hologram in the hologram plane in **a** which is propagated to the image plane **b** applying one wavelength. In **c**, the reconstruction process is performed with consideration of the wavelength, but only the amplitudes of the propagated waves are added in contrast to **d**, where a complex addition is utilised according to (4.9)

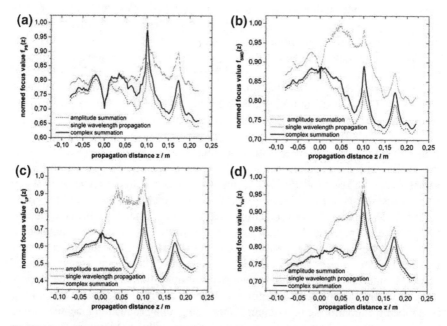

Fig. 4.14 Amplitude focus values $f(z)$ of different propagation methods as a function of the propagation distance z: **a** logarithmic power spectrum (PS), **b** Sum Modulus Difference (SMD), **c** Laplace (LP) and **d** Variance (Var)

Conclusion: The consideration of the spectral width of partial coherent light sources such as LEDs allows a further improvement of the quality of reconstructed and propagated holograms. Since the maxima of the focus values are more sharp, the robustness of the focus algorithm in finding the position of highest image sharpness is improved.

4.3.5 Tayloring of the Coherence Length

For calculation of the correlation and thus effective coherence length, which is used for analysis of e.g. topographies of specimens, a theoretical approach is demonstrated and experimentally verified.

Thus, this section deals with the development of a digital holographic microscope (DHM) which operates with low coherent light sources and variable coherence length that can be tuned via optical filters. This offers the possibility to adapt the effective coherence length during a measurement in order to adjust optimal parameters for a requested specific application. Especially when measuring optical components with multiple layers or stitched parts of a sample with different peak-to-valley (PV) values, this approach allows to suppress multiple reflections and stray-light. Thereby, the features of DHM enable a subsequent focus and thus a focusing of different sub-areas of a measured field which offers the possibility of faster measurements in a stitching mode over the whole positioning volume of a nano positioning system (e.g. NMM-1, SIOS GmbH), which is demonstrated in Sect. 4.8.

The application of superluminescence diodes (SLDs) as a low coherent light source not only requests to match the object and reference path length of an interferometer. Furthermore, the setup has to be compensated for dispersion [37, 67]. In those works, the calculation of the effective coherence length due to optical glass components in the experimental setup has been considered with approximations. By applying theoretical elements of optical coherence tomography, a comprehensive and more accurate formulation describing the interference of a set of different Gaussian shaped spectra can be found. The basic concept is the definition of a combined standard deviation and a combined centre frequency of the interfering spectra. In general, the approach can be generalised to arbitrary by decomposing such a spectrum into a set of Gaussian functions. Like in an OCT system, axial resolution - besides the wavelength - depends primarily on the coherence length of the light source [48]. In contrast, the lateral resolution depends on the numerical aperture of the optical system. Following these principles, considerable effort to improve the axial resolution of OCT by utilizing improved light sources with broader bandwidths to generate greater axial resolution and thus sharper cross correlation function has been expended. However, the assumed benefit of this is only completely realised if the spectra of the two interferometer arms of the interferometer are identical and not significantly altered by interaction with the optical system or the sample. The experimental investigations from an actual interferometer equipped with a broadband source may show an axial resolution significantly different from theoretical expectations in which identical spectra are supposed. In a low coherent digital holographic microscopy setup, a broad-band optical beam illuminates two separate pathways; the reference arm and the object or sample arm (see Fig. 4.7). Unlike many interferometers, such as in gravitational wave detection where the two arms of the interferometer are normally designed identically, the two arms of an OCT interferometer are usually different. This is especially the case for applications in biomedical imaging, but also for special coated optical components. In practice, each optical element, including biological tissue, has the potential to alter the spectrum of the light source beam, thus inducing differing spectra in each arm of the interferometer. The spectra of the light beam may show changes in amplitude, phase, centre wavelength and bandwidth. In order to predict those effects relatively precise in case of a DHM-setup, in the following, a general approach with appropriate simplifications is provided to describe the interference in a two beam interferometer in which the spectra returning from the two interferometer arms have experienced different filtering functions causing different centre wavelengths, amplitudes, bandwidths and phase.

In practice, it is convenient to assume that the local distribution of a non-Gaussian spectrum can be described in terms of a Gaussian function. Thus, a non-Gaussian distribution can be partitioned into a linear combination of Gaussian functions [70] in this more general approach:

$$\Psi(\omega, t) = \sum_{i}^{n} a_i e^{-\frac{(\omega-\omega_i)^2}{4\sigma_i^2}} e^{-j(\omega t - \phi_i)} \tag{4.16}$$

The corresponding intensity then yields:

$$I(\omega) = \Psi(\omega, t) \cdot \Psi^*(\omega, t) = \sum_{i=j}^{n} a_i^2 e^{-\frac{(\omega-\omega_i)^2}{2\sigma_i^2}}$$

$$+ \sum_{i \neq j} a_i a_j e^{-\frac{(\omega-\omega_i)^2}{4\sigma_i^2} - \frac{(\omega-\omega_j)^2}{4\sigma_j^2}} e^{-\Delta\phi_{ij}} \tag{4.17}$$

where a_i, σ_i, ω_i, $\Delta\phi_{ij}$ denote the amplitude, standard deviation, centre frequency of the ith Gaussian function and the phase shift between ith and jth parts. The subscript j represents the jth component of the complex conjugate wave. The first and second summations in (4.17) refer to the constant and interference parts respectively. The following analysis is derived using two Gaussian distributions, but can be generalised utilizing this assumption. It is often misleading, especially in high-resolution OCT, to assume that the spectra returning from the sample and reference arms of an interferometer are identical to the spectrum of the light source. To model the interferogram and derive an expression for an effective coherence length, we assume that the altered interfering fields have Gaussian spectral distributions, and the centre frequencies of the reference and sample fields are ω_1 and ω_2 respectively. The standard deviations of the fields are σ_1 and σ_2, respectively. Based on this asymmetrical Gaussian distribution assumption, the correlation function in the interference term should be a cross correlation function rather than an autocorrelation function. The interference term of the two beam interferometer can be written in the frequency domain as,

$$S(\omega, \Delta\phi) = Ae^{-\frac{(\omega-\omega_1)^2}{4\sigma_1^2} - \frac{(\omega-\omega_2)^2}{4\sigma_2^2}} e^{-j\Delta\phi} \tag{4.18}$$

where A and $\Delta\phi$ are the scale constant and the phase mismatch, while the $S(\omega, \Delta\phi)$ is the cross spectral density. The correlation function described above can be experimentally measured. However, an analytical formulation describing the phenomenon is useful for understanding the mechanisms and predicting the effects. In this study, we introduce a modified cross correlation function, which will clearly indicate the effect of spectral filtering. By defining a composite centre frequency and a composite standard deviation, (4.18) can be rewritten in the following,

$$S(\omega, \Delta\phi) = Ae^{-\frac{(\omega_1-\omega_2)^2}{4(\sigma_1^2+\sigma_2^2)}} e^{-\frac{(\omega-\bar{\omega})^2}{2\bar{\sigma}^2}} e^{-j\Delta\phi} \tag{4.19}$$

where the composite standard deviation $\bar{\sigma}$ and the composite center frequency $\bar{\omega}$ are respectively:

$$\bar{\sigma}^2 = 2\frac{\sigma_1^2\sigma_2^2}{\sigma_1^2+\sigma_2^2} \quad , \bar{\omega} = \frac{1}{2}\bar{\sigma}^2\left(\frac{\omega_1}{\sigma_1^2} + \frac{\omega_2}{\sigma_2^2}\right) \tag{4.20}$$

From these expressions, it can be noted that the Hadamard-product of two Gaussian functions is another Gaussian function. Thus, it can be shown that for the case in which the two interfering spectra are linear combinations of Gaussian distributions, the interference function will also be a linear combination of Gaussian distributions. The composite standard deviation $\bar{\sigma}$ only depends on the two initial standard deviations and is independent of the separation of the two spectra, while the composite centre frequency $\bar{\omega}$ depends not only on the separation of the two spectra, but also on their standard deviations. Importantly, the first exponential factor in (4.19) describes the attenuation of the correlation function due to difference between the two spectra. The phase mismatch in (4.19) contains a group delay time and a phase delay time, which determine the envelope of the interferogram and the frequency of the fringes. A continuation of the approach used above may be used to provide a clear, while more general, view of the group delay, phase delay, and effective coherence length. Assuming the fields propagate in two different dispersive media and the propagation constants are expanded at the different centre frequencies [71], the phases, $\phi_1(\omega)$ and $\phi_1(\omega)$, accumulated in double-pass as a function of frequency for the reference and sample fields, and the phase mismatch $\Delta(\omega)$ can be expressed as:

$$\phi_1(\omega) = \beta_1(\omega_1)(2l_1) + \beta_1'(\omega_1)(\omega - \omega_1)(2l_1) + \frac{1}{2}\beta_1''(\omega_1)(\omega - \omega_1)^2(2L) ,$$

$$\phi_2(\omega) = \beta_2(\omega_2)(2l_2) + \beta_2'(\omega_2)(\omega - \omega_2)(2l_2) + \frac{1}{2}\beta_2''(\omega_2)(\omega - \omega_2)^2(2L) , \tag{4.21}$$

$$\Delta\phi(\omega) = \phi_1(\omega) - \phi_2(\omega) , \tag{4.22}$$

where l_1 and l_2 are the propagation distances in the reference and sample paths, respectively, and L is the distance over which second order dispersion is encountered (assumed to be equal for both). For generality, we keep the dispersion coefficients β_1, β_1' and β_1'' for the first beam distinct from β_2, β_2' and β_2'' for the second beam. To derive group and phase delays, we substitute (4.19) into the last line of (4.22) and then rewrite it as:

$$\Delta(\omega) = \bar{\omega}\Delta\tau_p + \Delta\omega\bar{\sigma}^2 \left(\frac{\beta_1'(\omega_1)l_1}{\sigma_2^2} - \frac{\beta_2'(\omega_2)l_2}{\sigma_1^2} \right)$$
$$+ (\omega - \bar{\omega})\Delta\tau_g + L \left\{ \beta_1''(\omega_1)(\omega - \omega_1)^2 - \beta_2''(\omega_2)(\omega - \omega_2)^2 \right\} \tag{4.23}$$

where $\Delta\omega = \omega_1 - \omega2$, the phase delay $\Delta\tau_P$ and the group delay $\Delta\tau_g$ are defined as:

$$\Delta\tau_p = 2\frac{\beta_1(\omega_1)l_1 - \beta_2(\omega_2)l_2}{\bar{\omega}} \tag{4.24}$$

$$\Delta\tau_g = 2\beta_1'(\omega_1)l_1 - 2\beta_2'(\omega_2)l_2 \tag{4.25}$$

It should be noted that these equations simplify to the familiar formula used for OCT calculations in the case that $\beta_1(\omega_1) = \beta_2(\omega_2)$ and $\beta_1'(\omega_1) = \beta_2'(\omega_2)$ [72]. Through advantageous grouping and redefinition, (4.23) may be written in the more convenient form:

$$\Delta\phi(\omega) = \Delta\beta'' L \cdot (\omega - \bar{\omega})^2 + (\omega - \bar{\omega}) \cdot \Delta\tau_g' + \bar{\omega}\Delta\tau_p' \tag{4.26}$$

where $\Delta\beta'' = \beta_1''(\omega_1) - \beta_2''(\omega_2)$ is the mismatch of the group velocity dispersion (GVD), and the effective phase delay $\Delta\tau_p'$ and the effective group delay $\Delta\tau_g'$ are defined by:

$$\Delta\tau_p' = \Delta\tau_p + 2L \cdot [\Delta\beta''(\bar{\omega} - \omega_1) - \beta_2''(\omega_2) \cdot \Delta\omega] - \frac{L}{\bar{\omega}}[\Delta\beta''\bar{\omega}^2 - \beta_1''(\omega_1)\omega_1^2$$
$$+ \beta_2''(\omega_2)\omega_2^2] + \frac{\Delta\omega}{\omega}\bar{\sigma}^2 \cdot \left(\frac{\beta_1'(\omega_1)l_1}{\sigma_2^2} - \frac{\beta_2'(\omega_2)l_2}{\sigma_1^2} \right)$$
$$\tag{4.27}$$

$$\Delta\tau_g' = \Delta\tau_g + 2L \cdot [\Delta\beta''(\bar{\omega} - \omega_1) - \beta_2''(\omega_2) \cdot \Delta\omega] \tag{4.28}$$

Equations (4.27) and (4.28) indicate that both the phase and group delays are affected by not only the unequal spectral distributions, but also the mismatch of the GVD. It can be seen that (4.27) and (4.28) degenerate under simplifying assumptions, for instance, $\Delta\tau_p' = \Delta\tau_P$ and $\Delta\tau_g' = \Delta\tau_g$ when $\omega_1 = \omega_2$ and $\Delta\beta'' = 0$. Since in general, the media in the two paths are different, $\Delta\beta''$ is not necessarily zero for $\omega_1 = \omega_2$. Finally, the cross correlation function of an OCT interferometer interrogating dispersive media can be written in a more detailed form. Substituting (4.26) into

(4.17) and assuming scale constant $A = 1$, the final formula of the cross spectral density is obtained:

$$S(\omega, \Delta\phi) = e^{-\frac{(\omega_1 - \omega_2)^2}{4(\sigma_1^2 + \sigma_2^2)}} \exp\left(-j\bar{\omega}\Delta\tau'_p - j(\omega - \bar{\omega})\Delta\tau'_g - \left(\frac{1}{2\bar{\sigma}^2} + j\Delta\beta'' L\right)(\omega - \bar{\omega})^2\right) \tag{4.29}$$

The form of (4.29) is similar to the cross spectral density of two waves with identical spectral distributions described, for example, in equation (25) of chapter 2 of [72]. The difference here is the attenuation term at the beginning, and the use of effective phase and group delay, as defined above. Thus, the definition of the composite central frequency and effective phase and group delays generalizes previous expressions for two-beam interference in dispersive media [71, 72]. In the case of time domain OCT, a single point detector is used to collect the interference signal, so the signal current I should consist of the contributions of all the wavelengths under the interfering spectra. The integral is only performed over the last two factors in (4.29) because the first two factors are constants. After integrating (4.29) over frequency, we obtain the modified expression for the interferogram intensity:

$$I \propto \mathrm{real}\ [1 + \bar{\sigma}^4(\Delta\beta'' \cdot 2L)^2]^{-\frac{1}{4}} \cdot e^{-\frac{(\omega_1 - \omega_2)^2}{4(\sigma_1^2 + \sigma_2^2)}}$$
$$\cdot e^{-\frac{(\Delta\tau'_g)^2}{2\left(\frac{1}{\bar{\sigma}^2} + \bar{\sigma}^2 \cdot (\Delta\beta'' 2L)^2\right)}} e^{-j\left\{\bar{\omega}\cdot\Delta\tau'_p + \frac{1}{2}\tan^{-1}(\bar{\sigma}^2 \cdot \Delta\beta'' 2L) - \frac{(\Delta\tau'_g)^2 \bar{\sigma}^2 \cdot (\Delta\beta'' 2L)}{2\left(\frac{1}{\bar{\sigma}^2} + \bar{\sigma}^2 \cdot (\Delta\beta'' 2L)^2\right)}\right\}} \tag{4.30}$$

The third factor in (4.30) represents the envelope of the interferogram. From here, the value of the effective group delay corresponding to the position of the half of the maximum of the envelope, can be found to be:

$$\Delta\tau_{g\_\mathrm{half}} = \sqrt{1 + \bar{\sigma}^4(\Delta\beta'' \cdot 2L)^2}\frac{\sqrt{2\ln(2)}}{\bar{\sigma}} = \frac{1}{2}\mathrm{FWHM} \tag{4.31}$$

Equation (4.31) indicates that second order dispersion mismatch $\Delta\beta''$ extends the envelope. Assuming that GVD is matched in the two arms of the interferometer, i.e. $\Delta\beta'' = 0$, we can significantly simplify (4.30) to:

$$I \propto \mathrm{real}\left\{e^{-\frac{(\omega_1 - \omega_2)^2}{4(\sigma_1^2 + \sigma_2^2)}} e^{-\frac{\bar{\sigma}^2(\Delta\tau'_g)^2}{2}} e^{-j\bar{\omega}\Delta\tau'_p}\right. \tag{4.32}$$

Equations (4.30) and (4.32) are similar in form to previous descriptions of interferometric signal, e.g. [72], but are more general. Equations (4.30) and (4.32) reveal the attenuation of fringe visibility due to the difference between the spectral distributions of the two interfering fields and the effect on the group and phase delays of the two fields propagating in different dispersive media. Further, substituting (4.19) and (4.20) into (4.31) and assuming $\Delta\beta'' = 0$, we obtain the effective coherence length

l'_c as a function of the two individual field distributions, or of the coherence lengths l_{c1} and l_{c2} of the reference and sample fields,

$$l'_c = \frac{2\ln(2)}{\pi}\sqrt{\frac{1}{2}\left[\left(\frac{\lambda_1^2}{\Delta\lambda_1}\right)^2 + \left(\frac{\lambda_2^2}{\Delta\lambda_2}\right)^2\right]} = \sqrt{\frac{(l_{c1})^2 + (l_{c2})^2}{2}} \qquad (4.33)$$

The coherence length will be longer if $\Delta\beta'' \neq 0$ according to (4.31).

Experimental Investigations

In order to verify the executed calculations, several experimental investigations have been performed. For the experiments, SLDs with the centre wavelengths of 864.53 nm ($\sigma = 10.43$ nm), 760.33 nm ($\sigma = 9.44$ nm) and 767.16 nm ($\sigma = 7.50$ nm) have been utilised. To detect the predicted effects with unequal dispersion in the interferometer, the coherence length has first been additionally broadened by N-BK7 glass slides before entering the beam splitter. Furthermore, spatial filtered infrared LED light has been investigated with a wavelength of 850 nm (bandwidth: 32 nm) in combination with a spectral filter in one interferometer arm (centered at 871.6 nm, bandwidth: 6.7 nm). The detected signal in the interferometer with and without the filter are shown in Fig. 4.15. In the right figure, the ratio is defined as 32 nm (at 850 nm) divided by a variable spectral width. Dashed line: two spectra are identical and the width variations are the same for both spectra. Solid line: one of the two spectral widths is fixed at 32 nm and the other varies at the centre wavelength 871 nm.

This broadening of the interference function and therefore the resultant axial resolution degradation calculated from the formulae presented here mainly agrees with the experimental results according to Fig. 4.15. Thus it is shown that the theoretical approximations are basically sufficient for the applied experimental setups. It has to be considered that the contrast may decrease significantly when applying a strong mismatch of both interfering waves regarding the spectra.

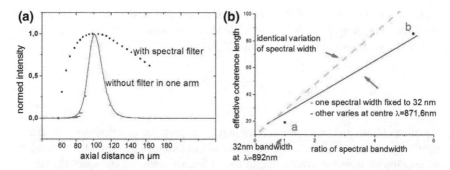

Fig. 4.15 a Illustration of measured intensity corresponding to the cross correlation function S with spectral filter (dashed) and without filter in one interferometer arm. Right: Plot of the effective coherence length as function of the spectral bandwidth ratio for two investigations. **a**: no filter, cross indicates the measurement of 22 μm, while the calculation is 24 μm; **b** with filter: down triangle indicates the measurement of 90 μm, while the calculation is 87 μm

Conclusion: It has been experimentally demonstrated, that the theoretical approach including simplifications to calculate the effective coherence length is done with sufficient accuracy. The investigations show, that with the applications of filters, an efficient adjustment of the available fringe number may be performed.

4.4 Error Compensation of SPM in a Nano-positioning Reference System

A variety of SPM (Scanning Probe Microscopy) techniques allow measuring different local physical properties of the surface or micro and nano scaled objects under investigation. One of the key properties of interest in research of micro- and nano-optics is a high local resolution of diffractive as well as refractive optical elements. Nanopositioning systems as introduced in Sect. 3.6 allow versatile applications like measuring topographies of objects, imaging with microscopes in a stitching manner or direct laser writing by integrating separately or simultaneously specific modules into such a positioning system (see also Chap. 3).

By integrating an atomic force microscope into a coordinate measuring machine (here: NMM-1, SIOS) with sub-nanometre resolution and nanometre uncertainty, an unprecedented atomic-force microscope is realised which enables a simultaneous investigation of form and roughness of specimens with sizes up to 25 mm × 25 mm × 7 mm along x, y and z-axes respectively. The calibration of the AFM-sensor is performed with different silicon gratings. Each grating is specially designed for investigating the lateral and axial resolution as well as the accuracy in slope measurement. Different modes of scanning strategies have been analysed and error compensated for micro and nanostructured optical components with a surface diameter of up to 25 mm.

Integration of an AFM-sensor into the NMM-1 The application of an atomic force microscope on the NMM shown in Fig. 4.16 opens up prospect for long range AFM scans in the full positioning range of the stage. A programmed NMM stage control loop holds the AFM system in the middle of its measuring range. The AFM head hysteresis can be neglected. In that way, the main surface slope is measured by the interferometers of the NMM stage and only fine structures by the AFM itself. Another advantage of the AFM application on the NMM is the possibility of correlated measurements on the surface in the nanometre range without stitching. This allows the measurements of additional artefacts features like angular relationship of structures or pitch distance with high precision. Uncertainties of pitch measurements of less than 0.1 nm are attainable. The utilised atomic force microscope (AFM, type: *DME DualScope DS 95-50*, Fig. 4.16b) offers a scanning volume of 50 μm ×50 μm ×2.7 μm. The AFM-head contains a piezo-tube scanner for the movement of the cantilever in x- and y-direction and an additional piezo-stack for movements in z-direction. The AFM-sensor is fixed with an adapter made of *Invar* on the *Zerodur*-frame of the NMM (Fig. 4.16a). It is adjusted in such a way, that the

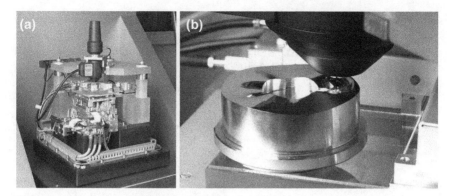

Fig. 4.16 **a** Setup of the nano-positioning and nano-measuring machine NMM-1 with a metrolog-ical frame made of Zerodur and a mounted AFM-sensor in the centre of the stage, **b** measurement of a micro-structured optical surface with an AFM-sensor

intersection point of the three (virtually extended) laser beams measuring the position of the comer mirror of the stage is identical with the position of the cantilever probe-tip. This set-up is in compliance with the Abbe-principle in all axes and therefore, this minimises measurement errors due to a tilt of the stage. The AFM-electronics offers two electrical outputs: the interferometer signal and a P-I feedback signal. The latter is a measure for the displacement of the piezo in the z-axis, because it is directly proportional to the applied voltage to the piezo actuator. This signal is fed into the NMM to calibrate the sensor and calculate the exact position of the cantilever during measurements.

Different operation modes of the system consisting of AFM and NMM have been examined and two of them selected for further investigations:

- **Mode 1**: the NMM moves the sample along the x- and y-axis to provide the scan motion, while the z-axis is kept constant. The inherent piezo-tube of the AFM-sensor moves the cantilever only in z-direction to follow the surface of the sample. While measuring, the position data of the moving stage (x, y, and z) and the cantilever (P-I-signal from the AFM) are recorded by the NMM control hardware. The profile information of the sample in the z-direction is calculated from the P-I feedback and the position data of the NMM (z-axis).
- **Mode 2**: the NMM moves the sample along the x- and y-axis to provide the scan motion. In difference to measuring mode 1 the z-axis is not kept constant. The NMM uses the P-I-signal of the AFM electronics as feedback to move the stage in order to follow the general surface structure of the sample. Since the NMM is too slow to follow smallest structures of the surface at a reasonable scanning speed, this subdivision is necessary to ensure optimal performance and operate both z-drives in a range with best characteristics. The advantage of his method is that the scanning range in the z-axis is not limited to $2\,\mu m$ any more. The position data of the moving stage and the cantilever (P-I-signal from AFM) are recorded.

4.4.1   Calibration and Error Compensation Methods

Demands

Before utilising the AFM-head in combination with the NMM, it has to be calibrated. Lateral standards such as 1D and 2D gratings are widely used to calibrate the magnification and to characterise the image distortions of the xy-plane of all kinds of microscopes. Methods such as metrological SPM, metrological SEM and optical diffractometry have been developed for the calibration of lateral standards.

At present, grating calibrations mainly focus on the determination of the mean pitch of the grating structures. However, its uniformity and rectangularity (for a 2D grating) also need to be calibrated. The uniformity determines how well a grating can be used as a transfer standard. The rectangularity of 2D gratings can, for example, be applied for the calibration of the rectangularity of other coordinate systems. While the measurement strategy and data evaluation method for step height measurements are almost mature, they are still under development for lateral measurements. Comprehensive measurements are performed to test the performance and accuracy of the system. Of highest importance are investigations of the z-axis of the AFM-head since it is the only axis which is not directly measured by the interferometric systems of the NMM but depends also on the accuracy of the AFM-head. The measurement errors of the other axis are mainly determined by drift and analysed in Sect. 4.4.1. For calibration of the axial and lateral resolution, standardised AFM calibration test charts are utilised (manufacturer: MikroMasch). Fourier transform (FFT-FT) method with measurement strategies for both 1D and 2D gratings have been proposed [73].

Nano- and microroughness Here, nano-roughness investigations of superfine surfaces with Ra values down to less than a nanometre are necessary and increasingly demanded. Such applications arise, e.g., from optical and X-ray mirrors. Depending on the surface characteristics, an AFM is usually an excellent tool for performing nano-roughness measurements. However, for measuring very fine surfaces with R_a values only at the sub-nanometre level, the noise of SPM needs to be taken into account. But in general the influence of the non-linearity, hysteresis and creep behaviour of the piezo on the nano-roughness measurements has to be considered. Besides nano-roughness, micro-roughness measurement is widely required in the quality assurance of manufacturing processes. Using scanning probe microscopes (SPMs) for roughness measurements has advantages such as both a very high vertical resolution (down to less than a nanometre) and a high lateral resolution (down to 10 nm or smaller). In contrast, optical methods have limited lateral resolution due to the diffraction limit of light, whereas stylus methods have limited lateral resolution due to the limited radius of the stylus tip (typically 2–5 μm). Additionally, SPM methods are able to measure surfaces in a number of detection modes: contact, intermittent contact and non-contact. The measurement force between tip and sample is low during the SPM measurement and, even in contact mode, reaches only a few nano-newtons. This prevents scratching of the measured surface during the measurement procedure. In contrast, stylus profilometers may cause scratches on soft material surfaces.

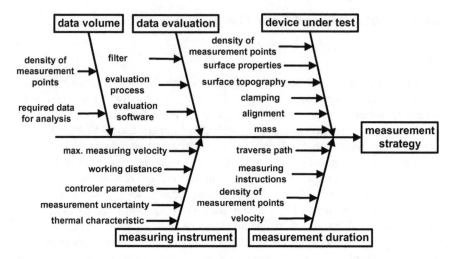

Fig. 4.17   Cause and effect diagram for nano-scale measurements with the NMM-1

Commercially available SPMs are less suitable for micro-roughness measurements of bigger surfaces due to the fact that they are not capable of a scan range of $7\lambda_c$ (λ_c is the cut-off wavelength, required by the standard roughness evaluation method according to *ISO 4288*). Fortunately the LR-SPM has a large measurement range of up to 25×25 mm^2 in the xy-plane. Several investigations on micro-roughness measurements using the large range SPM are carried out and described in the following.

Measurement strategy The measurement strategy can influence the measurement uncertainty significantly. The consequences of inappropriate chosen measurement and probing strategies with regard to conventional coordinate measurement techniques are summarised in [74]. These have to be modified and extended for ultra precise CMMs especially concerning the volume of data which has to be handled and the orientation precision of the clamping. Additionally, aspects on specific sources of uncertainty for nano-metrology have to be considered and are modelled in [75]. For nano-scale measurements with the NMM-1, a cause-and-effect diagram is depicted in Fig. 4.17. In order to demonstrate the influence of the orientation of the scan to the slope of the specimen, experimental investigations with a AFM sensor are performed. The results are taken into consideration in a written program for scanning procedures in the NMM-1. As standards for the z-axis, the calibrated step heights TGZ02 and TGZ03 are applied. The analysis is performed following the guidelines of DIN EN ISO 5436-1 (2006). Furthermore, for the lateral calibration, a *TGX01* test chart is applied. The reproduction of slope is analysed with a *TGF11* test chart. The sharpness or state of the probe tip is quantified with a *TGG01* test chart.

Correction of vertical drift Since in conventional AFM imaging, the surface under test is scanned in a raster (Fig. 4.18), the measurement process suffers from instrumental drift. Especially an AFM integrated into a micro and nano CMM is influ-

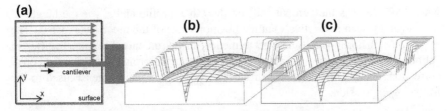

Fig. 4.18 **a** Raster of parallel scans on a surface under test with an AFM-sensor, **b** measured surface with scans along the fast scan direction (x-axes), **c** post processed data set with corrected vertical drift. The orthogonal red lines illustrate the set of profiles along the slow scan direction used for vertical levelling

enced by this error, whose extent usually has the highest impact on the topography in comparison to hysteresis and creep caused by the piezo-actuators of the scanner. Therefore, the measured topography is distorted and sometimes far from the ideal Cartesian condition of an equally distributed spacing. In consequence, the acquired data has to be considered as a sequence of parallel profiles along the scan direction (see Fig. 4.18) with an uncertain relative vertical position with regard to the fast scan direction, which means parallel to the AFM cantilever [76, 77]. As thermal drift is a time-dependent error influence, its effect within a single profile may be neglected. In contrast, high uncertainties occur in the slow scan direction since information on surface topography is superimposed with scanner system drift. Furthermore, possible dust contamination influences the alignment of the raster of performed scans. Dust particles can be picked up by the AFM tip, which is then offset vertically and suddenly discharged afterwards.

An error compensation can be applied in the framework of the post processing of data sets, as it is implemented in many commercially available software programs. But these algorithms utilise a polynomial fitting routine, performing a levelling of the sequence of profiles. The resulting smoothing effect along the direction perpendicular to the fast scan direction, designated as the slow scan direction, may have a negative effect on the measured topography. However, the slow scan direction is considered to be less reliable in comparison to the direction orthogonal hereto.

Thus, a method is applied which is compensating unequal spaced profiles by referencing to a set of profiles along the slow scan direction (Fig. 4.18). This approach is a simple and effective method for usual optical and thus smooth specimen. A more general method is performed by [76, 77], but since in the presented large scale AFM resolves in two and a half dimensions, an error correction with regard to the xy-plane is sufficient.

As depicted in Fig. 4.18b, c, the orthogonal scan along the slow axes only covers the inner part of the region of interest in order to reduce measurement time to a sufficient amount. Here, each single x-profile is vertically levelled and re-aligned according to (4.34) ($\forall k = 1, \ldots, m$):

$$\Delta z_k = \frac{\sum_{k=1}^{l} \left(z_{k,l}^{y} - z_{k,l}^{x} \right)}{L} \tag{4.34}$$

where Δk_z denotes the vertical shift on the kth x-profile and $z_{k,l}^x$ is the height of the x-scan (y-scan respectively) at the position (k, l) of the m × n data matrix. The reduction of distortions is usually performed by an instrument calibration on different calibration test charts for characterising the types of deformations. For lateral error correction, a classical model is introduced by [78] and can be generalised to the following expression:

$$\begin{bmatrix} x \\ y \end{bmatrix} = \begin{bmatrix} C_x & C_{xy} \\ 0 & C_y \end{bmatrix} \cdot \begin{bmatrix} x^* \\ y^* \end{bmatrix} \qquad M_{XY} = C \cdot M_{X^*Y^*} \qquad (4.35)$$

where x and y represent the physical coordinates related to each probed point, measured coordinates are denoted with a^* and C is the calibration matrix enabling the calculation of the metric coordinates M_{XY} through correction of the measured coordinates $M_{y^*x^*}$. Explicitly, the terms on the principal diagonal are scaling factors, while C_{xy} represents the linear coupling coefficient. Introducing nonlinearities and systematic superposition of drift, the concept can be modified in the following way [79]:

$$\begin{bmatrix} x \\ y \end{bmatrix} = \begin{bmatrix} C_x & C_{xy} \\ 0 & C_y \end{bmatrix} \cdot \begin{bmatrix} x^* \\ y^* \end{bmatrix} + \begin{bmatrix} a_x & b_x & & \\ & & a_y & b_y \end{bmatrix} \cdot \begin{bmatrix} x^{*2} \\ x^{*3} \\ y^{*2} \\ y^{*3} \end{bmatrix} + \begin{bmatrix} u_x \cdot t \\ u_y \cdot t \end{bmatrix} \qquad (4.36)$$

Nonlinearities are expressed with a third-order model, where a_x, b_x, a_y and b_y are the coefficients of the third-order polynomial function. The parameters u_x and u_y are the constituencies of the drift velocity vector; t is the time elapsed from the start of the scan to the measurement of the pixel lying in the position (x, y).

The offset between x-scan pixels and corresponding y-scan pixels is calculated by a differentiating model [79] applied to the two measurements:

$$\begin{bmatrix} \Delta x(x, y) \\ \Delta y(x, y) \end{bmatrix} = \begin{bmatrix} x \\ y \end{bmatrix}_{X\text{-scan}} - \begin{bmatrix} x \\ y \end{bmatrix}_{Y\text{-scan}}$$

$$= \left(\begin{bmatrix} C_x & C_{xy} \\ 0 & C_y \end{bmatrix}_{X\text{-scan}} - \begin{bmatrix} C_x & C_{xy} \\ 0 & C_y \end{bmatrix}_{Y\text{-scan}} \right) \cdot \begin{bmatrix} x^{*2} \\ y^{*2} \end{bmatrix}$$

$$+ \left(\begin{bmatrix} a_x & b_x & & \\ & & a_y & b_y \end{bmatrix}_{X\text{-scan}} - \begin{bmatrix} a_x & b_x & & \\ & & a_y & b_y \end{bmatrix}_{Y\text{-scan}} \right) \qquad (4.37)$$

$$\times \begin{bmatrix} x^{*2} \\ x^{*3} \\ y^{*2} \\ y^{*3} \end{bmatrix} + \begin{bmatrix} u_x \cdot t \\ u_y \cdot t \end{bmatrix}_{X\text{-scan}} - \begin{bmatrix} u_x \cdot t \\ u_y \cdot t \end{bmatrix}_{Y\text{-scan}}$$

In first order and with little error, the calibration matrix C can be considered constant; similarly also the mean drift velocity components can be considered as a constant when switching from the x-scan to the y-scan. With these approximations, (4.37) can be rewritten:

$$\begin{bmatrix} \Delta x(x, y) \\ \Delta y(x, y) \end{bmatrix} = \left(\begin{bmatrix} a_x & b_x \\ & \\ a_y & b_y \end{bmatrix}_{X\text{-scan}} - \begin{bmatrix} a_x & b_x \\ & \\ a_y & b_y \end{bmatrix}_{Y\text{-scan}} \right)$$

$$\times \begin{bmatrix} x^{*^2} \\ x^{*^3} \\ y^{*^2} \\ y^{*^3} \end{bmatrix} + \begin{bmatrix} u_x \cdot \Delta t(x, y) \\ u_y \cdot \Delta t(x, y) \end{bmatrix} \tag{4.38}$$

The parameter Δt denotes the time duration elapsing between the first passage (relative to the x-scan) and the second passage (relative to the y-scan) of the tip on (x, y) position in the topography matrix.

The offset between corresponding pixels described by (4.38) is the result of drift and non-linearities due to creep and hysteresis. As closed-loop scanners are only slightly influenced by creep and hysteresis, an additional approximation can be performed, which is only valid for metrological AFMs (4.39):

$$\begin{bmatrix} \Delta x(x, y) \\ \Delta y(x, y) \end{bmatrix} = \begin{bmatrix} u_x \cdot \Delta t(x, y) \\ u_y \cdot \Delta t(x, y) \end{bmatrix} \tag{4.39}$$

An ideal reproduction of the surface topography would have no errors due to pixel offsets if distortions, expressed by (4.38) and (4.39), were null. However, if the same deviations are less than half the pixel size in the topography, the errors have no influence on the result. Conversely, in relation to the dimension and to the geometry of the measured distribution of the z-values, appropriate results are also obtained if errors are higher.

Further considerations can contribute to the reduction of distortions. With regard to open-loop instruments, a considerable reduction can be attained both for hysteresis and creep deviations. Concerning the hysteresis, two approaches are possible: the first is an off-line correction, based on correction functions evaluated on calibration gratings; the second is an online correction based on optimisation of the scanner control. For the creep, it decreases logarithmically in time. The simplest way to substantially reduce creep effects is to repeat the tracking of the probe along the first line, before proceeding with the measurement of the x-scan and y-scan. Alternatively, one has to discard the affected zones (i.e., the margins where driving voltage inversion occurs) and perform reconstruction over zones where the creep effect is minor. Differently from hysteresis and creep, drift equally affects open-loop and closed-loop scanners. The main distortions are related to the x-scan: in fact the y-scan, which consists only of a few profiles, is very fast and is negligibly affected by drift. Since an off-line compensation of drift is hardly realisable, the maximum attention has to be paid to the measuring condition. A properly insulated environment and a sufficient time for thermal adaptation when starting the instrument can help in substantial reduction of drift velocity components. Some tests are performed on the open-loop AFM applied for this study. Scan settings were fixed to a scan speed of $5\,\mu m\,s^{-1}$ and a scan range $50 \times 50\,\mu m^2$ with a resolution of 2048×2048 pixels. Normally, a well-stabilised instrument can achieve drift velocities less than $0.02\,nm\,s^{-1}$. As already stated, these attentions can improve the quality of the reconstruction. Whenever these are not sufficient to reduce pixel offsets under acceptable levels, new reconstruction strategies are needed, based for example on larger y-scans and the use of a cross-correlation matching function for the compensation of residual distortions [76, 77].

4.4.2 Experimental Results

Measurements on Calibration Gratings

The *TGZ02* test chart for the axial calibration is depicted in Fig. 4.19. A ROI of $10\,\mu m \times 10\,\mu m$ is analysed according to DIN EN ISO 5436-1 (2006), where only the surface of an inner part of one step is taken into consideration for calculating characteristic parameters like the roughness. The results for the *TGZ02* are summarised in Table 4.1. Here, the measuring mode 1 in non-contact mode has the highest deviation with regard to the specification of $82.5 \pm 1.5\,nm$.

For a detection of nonlinearities and a calibration of the lateral resolution, a *TGX01* is applied. In Fig. 4.20a, b, the characteristic displacements of each square of the uncorrected and corrected data set is indicated with red arrows. The model which has to be fitted for compensating the non-linearities is of type $X = bx + cx^2 + dx^3$. The functions are shown in Fig. 4.21. This procedure has been performed for the whole measurement range of the NMM by an automated algorithm.

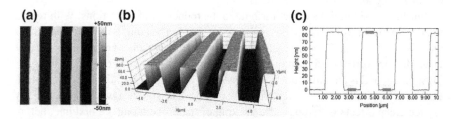

Fig. 4.19 **a** AFM-measurement of the calibration grating *TGZ02* with a ROI of $10\,\mu m \times 10\,\mu m$ in 2D-(**a**) and 3D-representation; **b** A profile with indicated regions in red used for the analysis according to DIN EN ISO 5436-1 (2006) in **c**

Table 4.1 Main results of the axial calibration with the *TGZ02* test chart

TGZ02	Contact-mode		Non-contact mode	
Specification: 82.5 ± 1.5	Step height(nm)	Rq(nm)	Step height(nm)	Rq(nm)
Meas. mode 1	83.306 ± 1.919	0.523	85.854 ± 0.808	0.881
Meas. mode 2	83.29 ± 1.765	0.638	82.707 ± 0.829	0.677

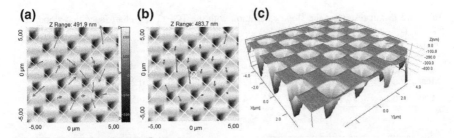

Fig. 4.20 Illustration of systematic errors caused by non-linearity: **a** non-corrected data set with characteristic red arrows indicating the amount of displacement of each square, **b** corrected data set with residual displacement vectors, **c** 3D representation of the specimen (*TGX01*)

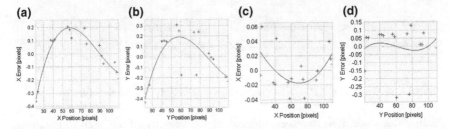

Fig. 4.21 Illustration of the residual error in x- and y-direction for compensating image distortion if applying the polynomial of order three $X = bx + cx^2 + dx^3$. The calculated coefficients are: **a** X: $b = 1.067228$, $c = -0.0008635016$, $d = 3.25961\text{E-06}$; **b** Y: $b = 1.074786$, $c = -0.0009537593$, $d = 3.568153\text{E-06}$; **c** X: $b = 0.9978405$, $c = 9.076962\text{E-06}$, $d = 6.215351\text{E-08}$; **d** Y: $b = 1.011720$, $c = -0.0002150203$, $d = 1.156946\text{E-06}$

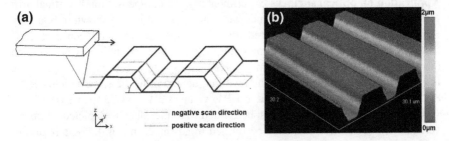

Fig. 4.22 **a** Illustration of the test procedure performed for calibration of the slope, **b** three dimensional representation of a ROI of the *TGF11* calibration test chart

In order to test the AFM-setup with regard to reproduction of slope, a *TGF11* calibration test chart is used (Fig. 4.22) with a slope of $\alpha = 4.733°$. For this calibration, several ROIs of $15\,\mu\text{m} \times 15\,\mu\text{m}$ were scanned with a velocity of $2.5\,\mu\text{m/s}$ along the fast scan direction as depicted in Fig. 4.22a in order to avoid errors due to high dynamic. The obtained results are summarised in Table 4.2. Table 4.2 reveals that especially the measurement mode 1 has significant deviations. Thus, measurement mode 2 has to be preferred in most measurements.

Table 4.2 Main results of the slope calibration with the TGF11 test chart (specified slope: $\alpha = 54.733°$)

TGF11	Contact mode		Non-contact	
	α_{left}	α_{right}	α_{left}	α_{right}
Meas. mode 1	47.492 ± 0.570	63.112 ± 9.468	45.818 ± 0.283	70.592 ± 2.7110^{-13}
Meas. mode 2	52.935 ± 0.832	55.584 ± 0.360	40.201 ± 0.591	54.561 ± 0.400

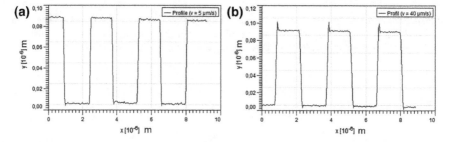

Fig. 4.23 Profile of a *TGZ02* test chart measured with a velocity of **a** 5 μm and **b** 40 μm

In order to demonstrate the influence of the velocity of the scan to measured topography of the specimen, further experimental investigations have been carried out (Fig. 4.23). The maximum error introduced has been investigated for different velocities and is considered in a program which locally reduces speed according to the measured slope to augment accuracy.

Application on optical surfaces Experimentally, the compensation of vertical drift is demonstrated in Fig. 4.23a on a microlens array which is also shown in Fig. 3.4a. Here, the upper part of a measured microlens array shown has been corrected, the lower part of the data set shows the original data. A statistical analysis on the obtained data has been performed for 20 lenses. The mean radius of curvature is $r_c = 349.582\,\mu m \pm 0.837\,\mu m$, the resulting deviation to the sphere yields $R_a = 8.354\,nm \pm 0.926\,nm$. The lenses have a PV-value of $PV = 4.732\,\mu m \pm 0.017\,\mu m$.

In Fig. 4.25c, the residual of a sphere fitted to a microlens is depicted. Here, no vertical error correction has been performed in advance. In Fig. 4.25a, a region of interest (ROI) of $500 \times 500\,\mu m^2$ of the same microlens array as shown in Fig. 3.4a and Fig. 4.24a is depicted. To demonstrate the applicability of the large range nanomeasuring machine for testing coated optical surfaces, a high resolution 1951 USAF test chart has been measured. An excerpt is shown in Fig. 4.25b. Here, optical testing is limited by the diffraction limit. The roughness values of the chrome and the glass surface in the white rectangles in Fig. 4.25b are $R_a = (4.8 \pm 0.4)\,nm$ and $R_a = (7.0 \pm 0.3)\,nm$ respectively.

Finally, to demonstrate the capability of nanoprecision measurements in the full lateral range which is offered by the NMM-1, a sphere normal with a diameter of 25 mm has been scanned with the AFM-sensor. Thus, a large scale AFM is not only applicable for micro-electro-mechanical systems (MEMS), but is also an ideal tool for quality testing of micro optics and optical test charts.

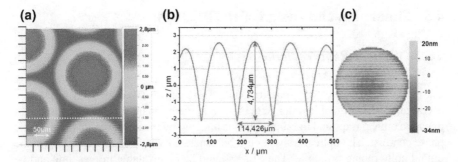

Fig. 4.24 a AFM-measurement of a microlens array with vertical drift compensation in the upper part of the data matrix; **b** profile of a data set along the vertex of a line of microlenses with indicated PV-value and diameter; **c** residual of a sphere fitted to one microlens to evaluate the quality

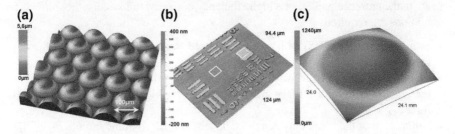

Fig. 4.25 a ROI of $500 \times 500\,\mu\text{m}^2$ of the same microlens array as in Figs. 3.4a and 4.19a; **b** scanned surface of a 1951 USAF high resolution test chart: testing of chrome coatings on glass. The roughness values of the chrome and the glass surface are $R_a = 4.8\,\text{nm} \pm 0.4\,\text{nm}$ and $R_a = 7.0\,\text{nm} \pm 0.3\,\text{nm}$; **c** utilisation of the full lateral scan range of $25 \times 25\,\text{mm}^2$ offered by the NMM-1 with a sphere normal used for optical calibration

Conclusion The realised large scale AFM-sensor has been calibrated with regard to the axial and lateral resolution, as well as concerning the slope measurement. The results of the calibration procedure are considered in a program for calculating optimised scan directions. The concept of drift compensation is also successfully applied to the large scale AFM in such a way, that the setup is considered to be an adequate reference measurement system and tool testing nano and micro sized objects with a diameter up to 25 mm. Performed measurements may therefore also serve for calibration of other less accurate systems.

4.5 Simulation and Design of HOT Setups

4.5.1 System Requirements

The digital holographic, SLM-based modular system to be developed shall to enhance a commercially available microscope by two functionalities: the generation of dynamic optical traps and the simultaneous imaging by means of digital holographic phase contrast. The optical traps should be generated with a SLM according to the digital holographic principle. The SLM should be controllable in real time and be able to change the phase of the reflected light. The phase pattern should be projected onto the entrance pupil (EP) of the microscope objective (MO). Since it is not directly accessible, the camera output of the microscope is used to transfer the modulated beam to the entrance pupil. For a high efficiency of the overall system, the following conditions are required:

- as many pixels of the SLM have to be imaged to the EP of the MO, so that the reconstruction of the hologram is as complete as possible
- illumination of the SLM with a homogeneous radiation, especially since the marginal rays in the EP make a disproportionately high contribution to the trap force
- collimation of the beam between the tube lens and EP to fit the infinite ray of the used microscope
- the reconstruction of the individual traps should offer a diffraction-limited behaviour over the entire field (maximal optical path length difference (OPD) $\leq \lambda/10$)
- the coupling lens has to image the internal diameter of the SLM precisely in the EP of the objective
- since multiple MOs are used, but the construction can not be constructed variable, an optimum imaging factor has to be found, which is a compromise for all the utilised MOs with different magnification
- the light efficiency of the module has to be as high as possible so that the laser power is sufficient for the generation of several traps
- for usual modern microscopes, a telescope optic has to provide parallel illumination to the infinity-corrected microscope objective since no tube lens is present

State of the art microscope objectives are designed for best imaging properties with the sample in the front of the focal plane. Thus the wave front is parallel after the objective facilitating an integration of further optics into the beam path with negligible change on aberration correction and focus [80]. In consequence, infinity corrected microscopes do not require lenses in the epifluorescence path to collimate the light. Therefore, in this case the position of the telescope can be chosen axially with high freedom in contrast to older microscopes with a fixed tube length [81].

4.5.2 Considerations on Optical Design

After reflection of the collimated light beam on the SLM, the beam has to be coupled into the imaging path of the microscope. Therefore, when considering especially infinity-corrected microscope objectives, the distance b_1 from the SLM to the telescope and that to the exit pupil b_2 (see Fig. 4.26 have to meet certain requirements. For an ideal function of the telescope, $f_1 + f_2$ is the distance of the telescope lenses.

Typically, the magnification of this 4f-telescope system is defined by $M = B'/B = f_2/f_1$ and is thus independent of the distances b_1 and b_2. The corresponding total length of the 4f-configuration is - with application of the precedent definition of M and the Gaussian lens relation for d_2 - given by:

$$L = f_1 + f_2 + b_1 + b_2 \qquad \text{with } b_2 = -b_1 M^2 + f_1 M(1 + M)$$
$$= (1 - M^2)b_1 + (1 + M)^2 f_1 , \tag{4.40}$$

Usually the SLM is positioned at the front focal plane of lens $L1$, which implies $f_1 = b_1$ and a sharp imaging to the plane at $f_2 = b_2$ behind the second lens L_2. Therefore the wave front is collimated by lens $L1$ and the overall length is then $L = 2f_1(1 + M)$. Alternative arrangements with increased or lower distance of the telescope to the SLM result in beam paths depicted in Fig. 4.27.

In Fig. 4.27a the entrance pupil EP is given by the lens diameter of $L1$, in (b) the distance of the EP to $L1$ may be significantly bigger than f_1. The case in (b) offers the advantage of a compact setup, since the SLM may be positioned directly in front of the first lens $L1$ as the magnification is independent on the SLM position (minimum of b_1 in (4.40)). The disadvantage is the loss of significant light power due to vignetting by the second lens of the telescope. The parameters L and b_2 as a function of b_1 are illustrated in Fig. 4.28a with $f_1 = 300$, $M = 0.5$. It is shown that the theoretical minimum value of L is at 675 mm at a SLM distance of nearly zero.

Since the beam diameter after reflection on the SLM usually has to be compressed by the telescope optics, the entrance pupil diameter EP_2 of lens $L2$ is usually smaller than that of lens $L2$, thus $EP_1 > EP_2$.

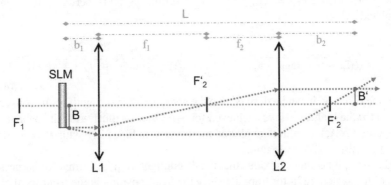

Fig. 4.26 Schematic sketch of the telescope unit typically utilised between SLM and microscope port with the lenses $L1$ and $L2$

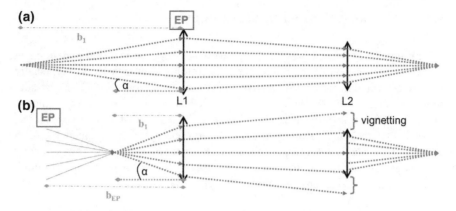

Fig. 4.27 Rays from the SLM traced through the telescope optic for different distances of the SLM: **a** $f_1 < b_1$, **b** $f_1 > b_1$

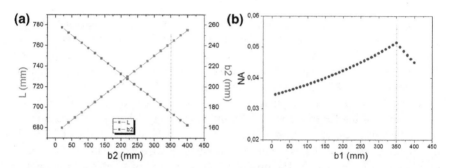

Fig. 4.28 Exemplarily results for illustration of dependence of total optical system length L and objective back focal length b_2 on the SLM-telescope distance b_1 in **a** as well as the dependence on the numerical aperture NA in **b** for $f_1 = 350\,\text{mm}$, $M = 0.5$, $\text{EP}_1 = \text{EP}_2 = 18\,\text{mm}$

In case (c) of Fig. 4.27b with $b_1 < f_1$, the lens is the system's aperture stop, leading to an EP on the left side of $L1$ as indicated in Fig. 4.27b. The distance from $L1$ to the EP is then given by $b_{\text{EP}(c)} = f_a(1 + M)/M$. The diameter of the entrance pupil yields then $\text{EP}_{(b)} = \text{EP}_{L2}/M$. The resulting system aperture $\sin \alpha$ can be calculated by:

$$\sin \alpha = \frac{\text{EP}/2}{\sqrt{(\text{EP}/2)^2 + (b_{\text{EP}} - b_1)^2}} \underset{\text{for case (b)}}{=} \frac{\text{EP}_{L2}/M}{\sqrt{(\text{EP}_{L2}/2M)^2 + (f_a(1 + M)/M - b_1)^2}}$$

(4.41)

From the equation it can be derived that the aperture increases with b_1. In both cases (a) and (b) the highest value is achieved for $f_1 = b_1$. For illustrating the behavior of the aperture NA, the values are plotted for this case in Fig. 4.28b with $f_1 = 350\,\text{mm}$, $M = 0.5$, $\text{EP}_1 = \text{EP}_2 = 18\,\text{mm}$.

Also here, it becomes obvious that the 4f-configuration is optimal for harnessing most of the system light for trapping since the SLM position in the front focal plane of $L1$ leads to a maximum of the NA.

In the following emphasis is put on experimental hardware limitations for the distances b_1 and b_2. When applying a - typically inverted - standard microscope, the minimum length b_2 may be in the range of $\approx$300–400 mm considering that the system is placed completely outside the microscope body. In the 4f-configuration, the total length of the telescope yields $L = 2(f_1 + f_2) = 2d_2(1 + M)/M$. Considering now that a typical SLM has a diameter in the range of [8, 20] mm (e.g. from Holoeye, BNS, Hamamatsu) and a standard oil immersion objective [2, 5] mm. This yields to a magnification in the range of $\approx$[0.1, 0.7] and to a length of $L = 1.5$ m with $M = 0.5$ and $b_2 = 350$ mm. Longer optical beam paths and thus low local wavefront slopes at the optical surfaces typically reduce aberrations [82], but since optical system aberrations can be compensated by the SLM as well (see also Sect. 4.6.2), short optical systems are preferred. For the design of a HOT-system it has to be taken into account that decreasing b_1 results in a reduction of b_2 by the relation $M = f_2/f_1$ and therefore this distance is posing specific limitations to the smallest size of the optical system.

Alternatively, transmittance SLMs offer the possibility of smallest configuration lengths [83], but have lower resolution and fill factors. Another possibility for reduced system space is the application of a beam splitter in front of the SLM with polarisers in front of the entrance and exit plane of the beam splitter cube [84]. Here two constraints - that of polarisation control and on-axis operation - can be met by the use of a non polarizing beam splitter. The drawback is the loss of typically 3/4 of the initial light power, which is often inadequate with regard to the available laser power.

The final experimental setup is therefore realised without beam splitter, but the angle of incidence of the SLM of typically 45° theoretically causes errors concerning the reconstruction of the hologram, as the distance of each SLM row and the objective changes in this configuration. This has been investigated by [85] and it has been demonstrated that when calculating a hologram with several trap spots with e.g. the Gerchberg–Saxton algorithm [86, 87], the effect on the resulting intensity distribution in the focal plane of the objective is marginal since the differences in the intensity peaks of traps, which have been initially equal, are varying only slightly in the range of $\approx$12%. Additional "ghost traps" may also be generated depending on the actual trap distribution. But typically only very few may arise in the range of up to 15% of the trap intensity of the intended ones. The simulation takes into account the actual phase-mostly modulation curve as well as the geometrical parameters of the SLM. Therefore, application of a SLM at high angles of incidence is causing no significant error and is the recommended approach for harnessing these devices in an efficient manner. Alternative results from similar studies are given in [88] which is general accordance with the conclusions given here.

4.5.3 Investigations on Experimental Optical System

The available components for the experimental setup mainly utilised in this work are a microscope, a SLM and a trap laser and are presented in the following in more detail:

Microscope: The used microscope is a confocal laser scanning and fluorescence microscope (Leica DMRXE/TCS SL). It allows in addition to transmitted light and reflected-light microscopy, the scanning of the object plane with a plurality of laser sources in order to generate three-dimensional fluorescence images. It is a microscope with a infinite optical design, i.e. the MO produces no real intermediate image, but images to infinity. Nevertheless, a hardly removable (tube) lens is visible in the camera port, which has to be taken into consideration in the final optical design of the HOT-module. An inverted microscope was not available for the works like it is often common in combination with optical tweezers. This has the disadvantage that samples in big observation containers, for example specific flow chambers for cell cultures, can not be imaged without preparation of a sample holder (object holder and cover glass).

SLM: The used SLM screen is a reflective LCoS micro display with a resolution of 1920×1080 pixels. The pixel pitch or distance is $8\,\mu m$, resulting in an active area of $15.36 \times 8.64\,mm^2$. The fill factor is 87%, and the screen can be irradiated with a maximum intensity of about $2\,W/cm^2$. The utilised version (PLUTO-NIR-2 from Holoeye) is optimised in efficiency and reduced noise interference for a wavelength between 750 and 950 nm. The average reflectivity is at 975 nm is 62%, and the maximum phase shift $2.4\,\pi$. A second LCOS-panel (PLUTO-VIS) is applicable for the wavelength range of 500–1000 nm.

Trap laser: The used laser is a butterfly-diode laser in a single-mode pigtail design, the fiber end is angled 8° to minimize back reflections. The wavelength is 975.5 nm for a spectrum range of 0.28 nm. The maximum output power is 330 mW (see Appendix B.1). A second laser with a wavelength of 1032 nm and $\approx 8\,W$ optical output power is also available.

Module for Optical Tweezers

For coupling the trap laser, the vertical camera port of the microscope is applied. In normal operation a camera can be attached by means of c-mount adapter at this point so that the real image generated by the tube lens is exactly at the level of the CCD sensor. This intermediate image is 69 mm above the flange (see Appendix B.1). Of the tube lens only the focal length of $f_{TL} = 200\,mm$ is known. Since the image for the camera goes through the same output a dichroic mirror is used, which lets pass the light in the visible range to >85% and reflects >90% of the light in the range of 750–1200 nm.

The beam modulated by the SLM needs to be imaged in the EP of the MO. However the EP is different for each MO. Data from the available MOs is listed in Table 4.3. The focal length can be calculated from the ratio of the tube lens focal length and the respective magnification. The diameter of the entrance pupil is then approximated by twice the value of the product of the numerical aperture and focal length: $d_{EP} = 2 \cdot NA \cdot f_{MO}$ [89]. To adjust the diameter of the modulated beam to the diameter of the EP, a coupling lens outside of the microscope is needed. The coupling lens and the tube lens form an afocal system which transfers the image of the SLM in the EP. For each lens a magnification factor $\alpha = d_{EP}/d_{SLM}$ can be calculated,

Table 4.3 Data of the microscope objectives in use with the working distance WD

Lens (Leica)	Magnification	NA	f_{MO} in mm	d_{EP} in mm	WD in mm	Immersion
HC PL Fluotar	50×	0.8	4.0	6.4	0.5	–
HCX PL APO	40×	1.25	5.0	12.5	0.1	Oil
HCX PL APO	63×	1.4	3.2	8.9	0.1	Oil
HCX PL APO	100×	1.4	2.0	5.6	0.09	Oil

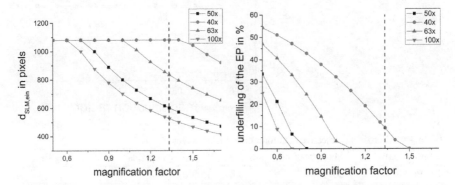

Fig. 4.29 Coupled SLM range (left) and underfilling of the EP (right) dependent on the magnification factor

whereby the inner diameter of the SLM fills the EP exactly. Here $d_{SLM} = 8.64$ mm is the diameter of the inner circle in the active area of the SLM screen. A system with a variably adjustable magnification factor could only be realised with increased effort, which is why an optimal magnification factor is determined which provides the best compromise for all useful lenses. Lenses with a high numerical aperture are preferable because the trap force increases with a higher incidence angle. Therefore, only the lenses with $NA \geq 0.8$ will be considered in the further investigations. Figure 4.29 (left) shows the dependencies between magnification factors and the diameter of the coupled in SLM image in pixels for the lenses. At a magnification factor of 0.6 all SLM pixels are coupled for all lenses. However, this causes especially for the 40× lens an underfilling of the EP of 50% (see Fig. 4.29(right)) and the important edge beams with a high angle of incidence are vignetted. Since the 63× and 100× objectives are mainly utilised, this drawback is not of high relevance.

With regard to the optimisation of the trap force, it is necessary to avoid an underfilling of the EP rather than having enough coupled pixels. For the same reason a further weighting of the effective numerical aperture of the lens is useful, which decreases with increased underfilling of the EP. The effective NA should be above 1, in order to realise sufficiently high trap forces. For the 40× lens this is the case from a magnification factor of 1.2 onwards (see Fig. 4.30).

A good compromise is a coupling lens with a focal length of 150 mm. Combining it with the tube lens results in a magnification factor of $1,\bar{3}$. This magnification of the SLM image leads to an underfilling of the EP only for the 40× lens, what

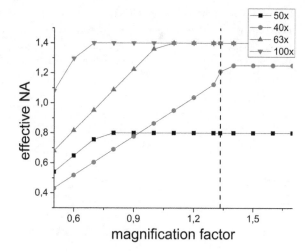

Fig. 4.30 Dependence of
the effective NA on the
magnification factor

Table 4.4 Microscope objective data for a magnification factor of $1,\bar{3}$

Lens (Leica)	Magnification	NA_{eff}	$d_{\text{SLM,ein}}$ in pixels	Underfilling EP (%)
HC PL Fluotar	50×	0.8	600	0
HCX PL APO	40×	1.15	1080	9.3
HCX PL APO	63×	1.4	833	0
HCX PL APO	100×	1.4	525	0

causes a slight reduction of the effective numerical aperture. For reasonable hologram reconstruction a minimum diameter of 600 pixels is recommended in literature
[90, p. 65]. This is almost achieved for the 100× lens (see Table 4.4).

The SLM should be illuminated with a homogeneous intensity distribution and a
plane wave front since intense edge beams contribute to higher trap forces. Taking
into account of the intensity profile of a laser beam perpendicular to the propagation
direction which is a Gaussian profile for lasers with TEM_{00}-mode, the beam illuminating the SLM has to be widened further than the diameter of the inner circle of the
SLM. The diameter of a light beam of Gaussian intensity profile is usually defined
at the point where the maximum intensity I_0 decreased to I_0/e^2 which corresponds
to 13.5% of I_0 (see Fig. B.17).

Thus, the marginal rays of the SLM have a sufficiently high intensity, the edge
pixels of the SLM should be still exposed to at least 50% of I_0. This value should not
be much higher, because otherwise too much power of the laser is not utilised. When
the radius of the inner circle in the active SLM area is $r = 4.32\,\text{mm}$ the intensity
should be 50% of $I_0(I(r)/I_0 = 0.5)$, which is achieved by a expanding the beam to
a diameter of approximately $2w = 14.7\,\text{mm}$:

$$I(r) = I_0 e^{\frac{-2r^2}{w^2}} \ , \quad w = \sqrt{\frac{-2r^2}{\ln(I(r)/I_0)}} \tag{4.42}$$

The selected optical fibre coupler collimates the laser beam to a diameter of 3.5 mm. To expand the beam to a diameter of approximately 14.7 mm, an afocal system with an expansion factor of 4.2 is required. To achieve this, a telescope consisting of two lenses with focal lengths of $f_1 = -50$ mm and $f_2 = 200$ mm is selected. Because of the use of an entrance lens with a negative focal length, this is denoted as a Galilean telescope, whose characteristics are a smaller installation space and no emergent focus point. The expansion factor of this system is only 4 - hence the beam is expanded to 14 mm and at the edge pixels of the SLM, the intensity of about 47% of I_0. This provides a sufficiently good approximation to the desired 50%.

From the (2.79) it becomes apparent that the trap force increases proportionally with the laser power. In the developed setup power losses occur at several surface planes. For this reason in the following the power losses are quantified as good as possible.

Power losses are induced by the polarisation filter of the SLM, the dichroic mirror, the beam expansion prior to the SLM and the overfilling of the entrance pupil of the MO. The transmittance of the polarizing filters is about 85% according to the manufacturer. The reflectivity of the dichroic mirror is >90% and of the SLM >99%. If not all SLM pixels are displayed in the EP of the microscope objective, power losses occur as well. The smallest EP is given by the 100× MO with a diameter of $d_{EP} = 5.6$ mm. Projected into the plane of the SLM, the diameter is only $d_{EP,SLM} = 5.6$ mm$/1.33 = 4.2$ mm due to the magnification factor. The diameter of the expanded light beam is $d_{Bel} = 14$ mm. For the calculation of the power loss of a Gaussian beam passing through a lens aperture with a radius a, the following formula is chosen [91, p. 362]:

$$P(a) = P_0 \frac{2}{\pi w^2} \int_{r=0}^{a} 2r\pi e^{-2r^2/w^2} dr = P_0 \left[1 - e^{-2a^2/w^2} \right]$$

$$P(d_{EP,SLM}) = P_0 \left[1 - e^{-2d_{EP,SLM}^2/d_{Bel}^2} \right]$$

(4.43)

For a maximum laser power of 330 mW, this results in a rest power of about 54 mW in the EP of the 100× MO, mainly due to the wide expansion of the beam. Taking into account the other known power losses, the maximal laser powers contributing to the trap forces are shown in Table 4.5.

In this configuration the laser power is sufficient to manipulate particles with a diameter >6 μm. But here the power losses of other components are not considered

Table 4.5 Maximal fraction of the maximum power 330 mW of the trap laser contributing to the trap force

Lens	Magnification	P_{max} in the EP (mW)	Ratio (%)
HC PL Fluotar	50×	52.4	15.9
HCX PL APO	40×	133.2	40.4
HCX PL APO	63×	91.1	27.6
HCX PL APO	100×	41.2	12.5

since the total laser power is projected in the EP and additional power losses at the polarisation filters as well as the SLM are not present. Furthermore in the developed system multiple particles have to be manipulated simultaneously. With each additional trap the available power of each spot is getting smaller, since the total power is chared with all spots. Therefore the maximum power of the first used, 330 mW laser [92] may be critical to manipulate many particles (approx. 5–10, depending on the size) simultaneously. However, the second laser with 8 W provides the required power [90, p. 150] for those applications, but has higher requirements with regard to laser safety.

Digital Holographic Phase Contrast Microscopy

Digital holographic phase contrast microscopy requires a further beam path of another laser in the visible range coupled into the microscope objective. The laser produces a hologram at the object plane, which is imaged on a camera. A continuous wave laser with the wavelength 532 nm is used, which has a high coherence length and a sufficiently high beam quality (see Appendix B.1). In order to generate coherent object and reference beams this laser is split into two beams using a fiber coupler with about 30 and 70% intensity division. The partial beam with the higher intensity is collimated and coupled into the transmission illumination port of the microscope. Through the microscope base the beam is finally transferred to the condenser which images it Koehler-like in the object plane. In the base of the microscope is usually a diffuser plate which is used for the homogenisation of a normal halogen light source. However, used together with a laser it induces an unwanted speckle pattern and therefore has to be removed.

The phase distribution of the transmitted laser light is varied by the specimen depending on the refractive index distribution. The MO and the tube lens magnify the image of the object plane, and finally generate a real intermediate image 69 mm above the flange. The beam emerging from the camera output is then superimposed with the reference beam by means of a beam splitter cube. Due to interference this gives rise to the actual hologram which is then imaged with a lens onto a CCD sensor. The tilting angle between the object beam and reference beam which is required for the "off-axis" setup is adjusted by means of the rotatable beam splitter cube.

Overall System

Figure 4.31 shows the basic setup of the overall system. The trap laser beam is collimated with a fibre coupler and expanded with a Galilean telescope consisting of two lenses (L1 and L2). For a "phase-only" operation of the SLM a linear polariser or a $\lambda/2$ retardation plate (P) are required which polarizes the laser beam linearly to the longitudinal axis of the LCoS screen and another linear polariser after the reflection on the SLM as an analyser (A) (see also Sect. 4.6). After the beam is reflected at a dichroic mirror (DM), the beam is coupled into the microscope beam path through the camera output. The lens system given by the coupling lens (L3) and the tube lens (TL) enlarges the image of the SLM in the entrance pupil of the microscope objective (MO), which finally Fourier transforms the wave in the object plane. The laser for digital holographic phase contrast microscopy is divided into object wave

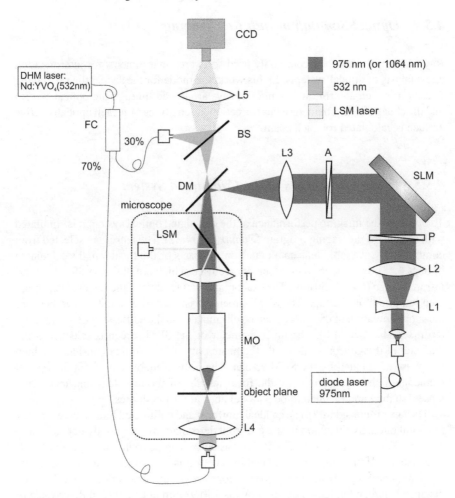

Fig. 4.31 Basic setup of the overall system. $L1, L2, \ldots$: lenses; P: polariser; SLM: spatial light modulator; A: analyser; DM: dichroic mirror; BS: beam splitter; FC: fibre coupler; CCD: camera; LSM: laser-scanning microscope; TL: tube lens; MO: microscope objective

(70%) and reference wave (30%). After collimation, the object wave is imaged by the condenser (L4) into the object plane and enlarged by means of the microscope objective (MO) and tube lens. After passing through the camera output, the light wave is transmitted through the dichroic mirror. Object and reference wave interfere within the beam splitter cube (BS) with a set tilt angle. The resulting hologram is imaged by the lens (L5) onto a CCD camera. Furthermore, the microscope allows confocal laser scanning measurement (LSM), wherein an internal laser scans the object plane. However, this mode can not be utilised simultaneously because in the applied microscope model the camera output must be closed temporarily for this operation.

4.5.4   Optical Simulation with Ray Tracing

Ray tracing programs are commonly used software environments for interpretation and analysis of optical systems. In this work, the mode for the sequential ray tracing is used in which an arbitrary number of rays are traced through the optical system and the change in the characteristics (direction, intensity, etc.) at each optically active surface is calculated for each beam.

4.5.5   Optical Properties of the Calculated System

To optimize the imaging performance of the calculated optical system it is simulated with a standard ray tracing program (Zemax, Fig. 4.32). The lenses are selected from commercially available lens catalogues on the basis of the focal lengths calculated in Sect. 4.5.3. Here achromatic doublets are used which are optimised for the near-infrared range (650–1050 nm). The exact manufacturer data of the selected lenses are then imported into Zemax. The SLM screen is assumed to be a flat mirror, because the shape deviation of the surface can be eliminated via the addressing of a correction term (see also Sect. 4.6). The inner diameter of the SLM screen is selected as the diameter of this mirror, so that only the beam portions are further considered,which are phase modulated by the SLM screen. The only available data from the optical elements of the microscope are the focal lengths of the tube lens and measuring lenses, so they are assumed to be paraxial aberration-free lenses.

The laser is considered to be an ideal, uniform and collimated light source, i.e. any temporal intensity fluctuations or wave front deformations are initially not taken into account. The overall design data of the lens editor can be found in Appendix A.2.1. The distance of the lenses is calculated by an optimisation function in Zemax so that the divergence angle behind the two telescopes gets a minimal value. The calculated distance between the lenses one and two is 140.93 mm at an angle of divergence of $2.5 \cdot 10^{-8}$ rad and between the lens 3 and the tube lens 344.63 mm at an angle of divergence of $6.0 \cdot 10^{-18}$ rad.

The achromatic lens L2 in Fig. 4.31 is oriented that way that the spherical wave front hits the flat surface of the achromat L2 first. In this arrangement an additional optical element occurs whereby the maximum path length of the light decreases by a factor of 25 (Fig. 4.33).

Since the marginal rays contribute most to the axial trap force, the beam is more expanded than the SLM surface for homogeneous illumination. In this way, laser power is lost, but due to the Gaussian intensity distribution, the intensity of the radiation at the edge of SLM pixels is higher. Figure 4.34 show the intensity profiles of the beams in front of the SLM and the microscope objective. It is calculated that the expansion in front of the SLM leads to a higher intensity of the marginal rays and that the telescope consisting of the coupling lens and the tube lens increase that section by a factor of $1.\bar{3}$.

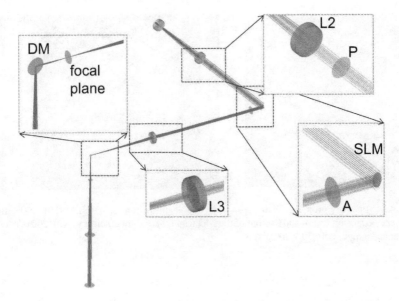

Fig. 4.32 Three-dimensional view of the ray path in the simulated system with partial enlargements. L2, L3: lenses, P: polarizer; SLM: spatial light modulator, A: analyzer, DS: dichroic mirror

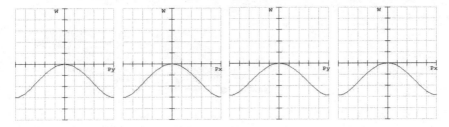

Fig. 4.33 Optical path length distribution (OPT) behind the beam expander for two different orientations of the second lens L2 at $\lambda = 976$ nm: Left: Y-scale $= \pm 0.5\lambda$. Right: Y-scale $= \pm 0.02\lambda$

The decisive factors for the quality of optical traps are the diffraction-limited characteristics of the system and the resulting wave front distortion in the object plane. The focal point in the object plane with a RMS Radius of 14 nm is substantially smaller than the radius of the Airy disc (0.63 μm) which is an indication for diffraction-limited performance. The wave front distortion in the object plane has a PV value of 0.007 wavelengths ($\hat{=}$6,825 nm) what is also sufficiently small (see Fig. 4.35). However, it has to be taken into consideration that the aberration free paraxial lens of the simulation is mainly responsible for this result and that significant deviations caused by aberrations have to be expected in the real system since the MO-design-data are not available by the manufacturers.

Tolerance Analysis and Sensitivity of Optical Setup

The used lenses and the mechanical supports are provided with certain production-related inaccuracies. Additional errors are induced on the assembly and adjustment of

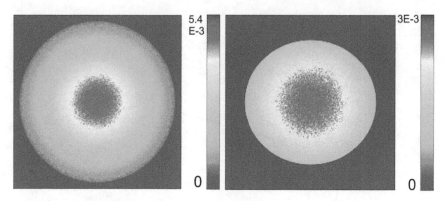

Fig. 4.34 Intensity profile in Watt per mm^2 of the beam in front of the SLM (left, 100% light efficiency (transmittance)) and in front of the MO (right, 58% light efficiency) with 200 mW laser power and a matrix of 330×200 pixels

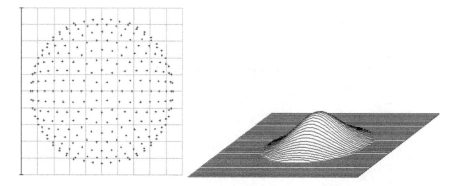

Fig. 4.35 Spot diagram (left, RMS radius: $0.014\,\mu m$; displayed area: $0.04 \times 0.04\,\mu m^2$) and three-dimensional wavefront deviation of the coupled trap laser in the focal plane of the microscope (right, $\lambda = 0.9750\,\mu m$; $PV = 0.0074\lambda$; $RMS = 0.0019\lambda$; Diameter $\approx 32\,\mu m$)

the setup and can not be completely eliminated. Thus the behaviour of the real optical system may differ from the calculated results. In order to quantify this influence, the impact of existing tolerances of optical components and errors in the adjustment are examined in more detail. Therefore, a Monte Carlo analysis in a simulation environment is applied. In this procedure, after setting the minimum and maximum tolerances regarding selected error parameters, they are varied stochastically. In the context of the stochastic analysis, the cases with best and worst configurations are shown, which could occur in the system. For the tolerancing, data available from the manufacturer of a corresponding optical component are applied. In case the variables are unknown the default values of the optical simulation software Zemax are adopted. Table 4.6 shows the utilised parameters.

The simulated worst case in the Monte Carlo analysis shows a clear distorted radiation distribution in the object plane with an RMS radius of $0.05\,\mu m$ (Fig. 4.36(left)).

Table 4.6 Used system tolerances

Parameter	Tolerance
Surface radius	0.2 mm
Thickness	0.15 mm
Decentering	0.1 mm
Tilt	0.1°
Surface irregularities	0.2 waves
Refractive index	0.001
Abbe number	1%

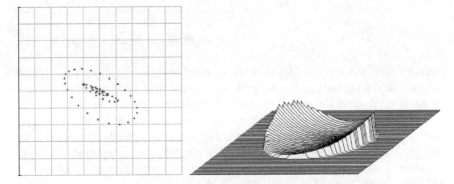

Fig. 4.36 Tolerancing results of the worst case scenario of the Monte Carlo analysis: spot diagram (left; RMS radius: 0.054; depicted area: $0.4 \times 0.4\,\mu m^2$) and three-dimensional wavefront deviation of the coupled trap laser in the focal plane of the microscope (right, $\lambda = 0.9750\,\mu m$; $PV = 0.0765\lambda$; $RMS = 0.0149\lambda$; Diameter $\approx 32\,\mu m$)

However, the system is still in the diffraction-limited range. The wave front deformation is increased by a factor of 10 to 0.07 wavelengths ($\hat{=}68.25$ nm) but is also still in the diffraction-limited range (see Fig. 4.36(right)).

To determine the adjustment sensitivity, the imaging performance of a system is examined whose beams are not accurately on the optical axis. Therefore, different wavefronts with varying angle of incidence are used. The resulting wavefront error, depending on the angle of incidence is shown in Fig. 4.37. It is demonstrated that the wavefront error rises as a function of the angle of incidence of the laser beam. The diffraction-limited behaviour disappears at an angle of incidence of $\approx 0.5°$. A precise adjustment of the optical axis is therefore essential to ensure sufficient quality of the optical traps.

4.5.6   Optomechanical Setup

The SLM manufacturer states that for optimal operation the screen has to be irradiated with the smallest possible angle of incidence since otherwise errors in the hologram

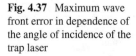

Fig. 4.37 Maximum wave front error in dependence of the angle of incidence of the trap laser

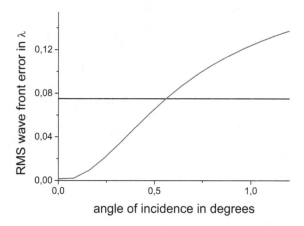

angle of incidence in degrees

reconstruction would occur. However, a very small angle would also require a much longer optical path, because this would lead to partial shading of the beam. Montes-Usategui et. al. have shown that the errors induced by a greater angle between the irradiated and the reflected beam are small and that even at an angle of 90° good results are obtained. If errors are unwanted they can be completely eliminated at the hologram calculation by considering the SLM tilt [93]. As a compromise between the induced error and compactness of the module, an angle between the beams of 45° is selected in front of and behind the SLM.

Due to cost and time constraints mostly commercially available standard components are used for the setup. The required focal lengths and distances of the lenses were calculated in the previous sections. Components of a 30 mm optical rail system are used as lens mounts and brackets. The brackets for the SLM screen and the associated driver unit were designed and manufactured as well as the casing for protection against laser radiation. Through displacement of the carriage, the centre of the SLM may be set to the optical axis. The base plate is connected to the SLM panel by three screws. Along the screws three elastically deformable plastic rings are placed. By tightening the screws the screen can be tilted about two axes to align the reflected beam in parallel to the optical axis of the module.

The overall system composed of microscope and developed module is shown in Figs. 4.38 and 4.39. Due to the weight and stability, the trap module attached to the microscope is held by a supporting rod. Because of the beam splitter cube, the dichroic mirror as well as the beam splitter, the CCD camera can no longer be installed on the designated place where a sharp real microscope image is projected by a tube lens 69 mm above the flange. For this reason, an additional lens is also required that transfers the image to the plane of the CCD chip. Because of the divergence of the image beam in combination with the diameter of the optomechanical components and the required space, partial vignetting of 10–15% of the image may occur due to the limitations of the applied microscope model and the restriction to easily available catalog lenses (see Fig. 4.40). Therefore, here a slightly reduced area from the center of the enlarged image can be imaged.

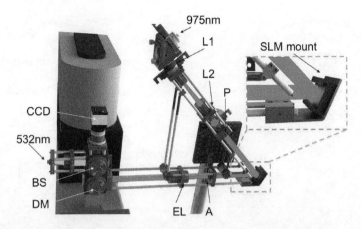

Fig. 4.38 CAD model of the overall developed system with the main components: fibre coupling of the trap laser (975 nm), beam expander consisting of the lenses $L1$ and $L2$, linear polariser (P), tilt- and shiftable SLM bracket, analyser (A), coupling lens (EL), dichroic mirror in a kinematic cube (DM), fiber coupling of the reference wave of the imaging laser (532 nm), beam splitter cube (BS) and a CCD or CMOS camera

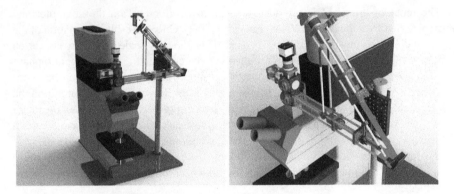

Fig. 4.39 More illustrations of the entire system from Fig. 4.38

4.6 Characterisation of the SLM

The used SLM is a "phase only" - SLM. It consists of a driving unit to which the actual display is connected via a flexible cable. The screen (PLUTO NIR-2) is a liquid crystal on silicon (LCoS) active matrix with a resolution of 1920×1080 pixels with an active area of $15.36\,\text{mm} \times 8.64\,\text{mm}$. Its efficiency is optimised for a wavelength of 700–1000 nm. In order to improve the quality of the modulated beam, the phase shift for the employed wavelength of 975 nm has to be calibrated. Furthermore, the production-related form error of the screen is measured and eliminated by a suitable superposition on the addressed phase patterns.

Fig. 4.40 Illustration of vignetting effects of the microscope image caused by a limited diameter of the opto-mechanical components of 30 mm and the additional required space between the real image and the CCD sensor

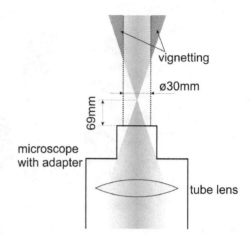

4.6.1   Calibration of Linear Phase Shift

The input of the SLM is provided through a monitor port on the computer and transmitted over the Digital Visual Interface (DVI). Only the 8-bit information of the green color channel of input signal is used. This 256 grey scale values are converted with a "data look-up table" (LUT), called gamma curve in the following, into phase values. Although the input signal consists of 8-bit information, the gamma curve has a length of 10 bits. The additional information is used by a correction circuit on the driver chip. The pixels of the SLM are addressed digitally and the values of the phase are generated by a pulse code modulation. Since the LC molecules have a certain viscosity, the addressing induces a flicker of a multiple of the sampling rate 60 Hz of the device. To reduce this flicker, it is possible to configure the device with a reduced sampling depth (number of bits) with the drawback of a reduced number of phase values.

By changing the gamma curve, it is possible to adjust the phase shift with the 256 grey scale values to the application and to the used wavelength. The application of optical tweezers requires a linear phase shift from 0 to 2π with the wavelength of 975 nm. The setup for characterisation of the phase shift of the SLM is shown in Fig. 4.41.

Using the experimental setup, the expanded laser beam is first linearly polarised and then split by a double pinhole into two partial beams. These are reflected on different regions of the SLM screen and analysed by another linear polariser. The two polarizing filters are aligned that way that their transmission axes are parallel to the longitudinal axis of the SLM screen. This setting allows a pure phase modulation with an amplitude which is as constant as possible. Subsequent, the two beams are superimposed by a lens system consisting of an achromatic lens with focal length of 150 mm and a microscope objective ($20\times$, $NA = 0.35$) and are displayed on a CCD chip. By tilting of the two partial beams, an interference pattern is formed by the superposition, as visualised in Fig. 4.41 (right).

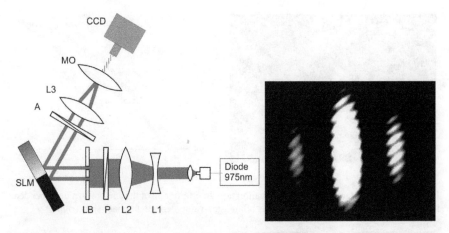

Fig. 4.41 Left: The laser beam is expanded by the telescope consisting of the lenses $L1$ and $L2$, polarised by the linear polariser (P), and divided in two individual beams by a pinhole (LB), reflected at the SLM and analysed by another linear polariser. The two beams are superimposed by the lens $L3$ and imaged onto a CCD camera through the microscope objective (MO). Right: Interference pattern of the superimposed beams

The spacing of the interference fringes is calculated with the formula:

$$\Delta x = \frac{\lambda}{2 \sin \Phi} \tag{4.44}$$

Thereby 2Φ is the angle between the tilted sub-beams. In the experimental setup, the distance between the CCD and lens $L3$ is about 380 mm. The distance between the collimated sub-beams is 5 mm. Thus Φ can be calculated with $\Phi = \arctan(2, 5/380)$ and at a wavelength of 975 nm this leads to a distance Δx of 74 μm. Multiplied by the MO magnification factor of 20×, this results in a Δx of about 1.4 mm on the CCD chip. The interference pattern within the (main) interference fringes is tilted by 45° and likely originated from back reflections within the CCD chip between the CCD and the cover glass. In order to eliminate them, the removal of the cover glass would be sufficient. But as it is not disturbing for the other measurements and thus this may be neglected.

If a split screen is now addressed on the SLM, the first half is constant black (value 0) and the other half is varied in 256 grey scale values, a variable relative phase shift of the two partial beams is generated. This causes a shift of the fringes of the interference pattern. A linear gamma curve provides a non-linear phase shift due to the nonlinear characteristic curve of the LC molecules. This is shown in Fig. 4.42(left), where the intensity pattern of a section of the interference image is on the x-axis and the 256 different grey values are on the y-axis. From the measured phase shift, a gamma curve can be calculated which enables the user to have a linear phase shift (see Fig. 4.43).

In addition to a linear phase shift of 2π, a grey value of independent, constant amplitude is of advantage for the application of optical tweezers. The amplitude

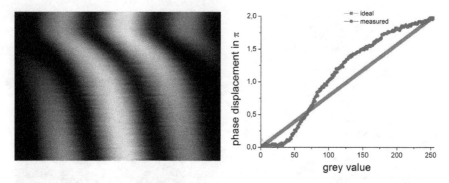

Fig. 4.42 Phase shift of the SLM before calibration: Interferogram of the 256 grey-value-dependent interference pattern (left) and the corresponding grey-value-dependent phase shift (right)

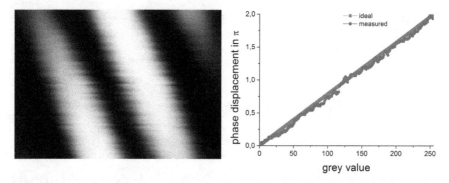

Fig. 4.43 Phase shift of the SLM after calibration: composite image of the 256 grey-value-dependent interference patterns (left) and the corresponding grey-value-dependent phase shift (right)

modulation is determined by a slightly modified setup. Therefore one of the two partial beams is blocked and the CCD camera is replaced by a photo-sensor. The photo-sensor detects the intensity of the nonblocked partial beam while an image is addressed on the SLM, which is changed in 256 grey levels. The measured intensity is normalised to the maximum value and is shown in Fig. 4.44. Across the entire grey-scale range, the maximum intensity varies by about 8%. Thus, the amplitude modulation of the SLM is within acceptable limits for the application of optical tweezers. The overall behaviour of the SLM is shown by the graph in Fig. 4.44. The deviation from the ideal characteristic is within a reasonable range.

4.6.2 Correction of the System Inherent Wave Front Aberrations

The SLM consists of a cover glass, followed by the active pixel array with a LC-molecular layer and with an aluminium coated silicate back plate (see Fig. 2.11). The

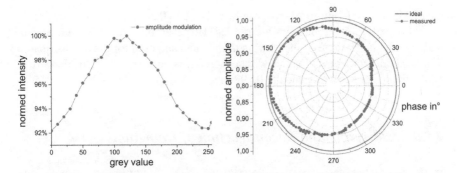

Fig. 4.44 Gray value dependent amplitude modulation normalised to the maximum intensity (left) and the locus (characteristic phase curve of the entire SLM behaviour (right)

production of this back plate is subject to certain tolerances, so that the quality of an optical flatness is generally not achieved. This induces additional aberrations to the coupled beam, which reduce the quality of the optical traps. The aberrations are wave front deformations caused by optical path differences. Since the system should be diffraction-limited in order to ensure the best possible trap quality, the shape error of the SLM panel has to be eliminated as far as possible.

To determine the form error of the SLM, the surface of the SLM panel was measured with a temporal phase-shifting Fizeau-type laser interferometer (*VeriFire AT+*, *Zygo*) with a repeatability of approx. <2 nm and form error of <15 nm.

The measured form error of the SLM panel is shown in Fig. 4.45(left). With a maximum PV value of 1.042 µm. For the correction the measured shape deviation at the wavelength of the trap laser is addressed inversely and displayed *modulo*(2π) on the SLM. For the wavelength of 975 nm, the grey-scale image is shown in Fig. 4.45(right). The phase hologram has to be superimposed with the holograms for optical traps for the subsequent application of optical tweezers, as long as the induced wave front error reduces the quality of the optical traps. By addressing the correction term on

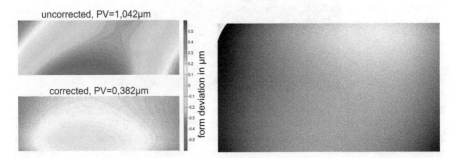

Fig. 4.45 Form deviation of the SLM panel with the power switched off (left, top) and with addressing of a correction image (left, bottom) and correction phase hologram calculated from the shape deviation for a laser wavelength of 975 nm (right)

the SLM, the maximum form error can be reduced to $0.382\,\mu\mathrm{m}$. For a more precise correction several iteration steps are needed. However, one must take into account that only the inner circle of the SLM is illuminated and used for the reconstruction. The maximum form deviation (i.e. without correction) in the inner circle is $0.486\,\mu\mathrm{m}$, and after the correction reduced to $0.192\,\mu\mathrm{m}$.

4.6.3   Addressing and Reconstruction of Holograms

First, with a simple setup, the functionality of an optimised SLM is demonstrated by hologram reconstructions. For safety reasons, the laser with a wavelength of 532 nm is used instead of the IR laser. The laser is widened with a fibre and collimated with a lens, illuminating the SLM panel on a kinematic unit, which is pivotable around three axes. Another lens images the modulated beam on a CCD camera (see Fig. 4.46).

To control the SLM an application software is utilised. This offers predefined holograms of simple signals that are shown on the SLM over the second display port of the graphics card. Furthermore, it allows to calculate and display the holograms of low resolution bitmaps (200×200 pixels) on the SLM.

Figure 4.47 shows the optical reconstruction of a hologram of the Fraunhofer IPT logo, displayed on a CCD camera. The zero order of diffraction is eliminated here by algorithms [94]. The quality of the demonstrated hologram is not optimal here since only simple lenses were used and the experimental setup was not optimised with respect to minimisation of aberrations. Furthermore, artefacts increase with more complex structured holograms. For focal spots, as it is necessary for the optical tweezer and the optimised expansion module, significantly less artefacts can be expected in the reconstruction.

Fig. 4.46 Basic setup of the optical hologram reconstruction consisting of the imaging laser (1) with attached fiber (2), a lens for collimation (3), an aperture (4), the SLM driver unit (5). SLM panel on a kinematic unit (6), a polariser (7) and an additional lens (8) for imaging the modulated beam onto a CCD camera (9)

Fig. 4.47 Optical reconstruction of the Fraunhofer IPT logo recorded with a CCD camera with zero order removal (left) and projected on a white screen and photographed with a digital camera with zero order diffraction spot in the center (right)

4.7 Determination of Optical Force and Trapping Stability

The possibilities opened up by digital holographic microscopy for complex manipulation and handling of particles especially in the lower micrometer range with optical tweezers has been demonstrated manifold. Contrary, applying HOTs for quantitative measurements is limited as the reproducibility and precision is confined by the spatial light modulator and its technique for displaying the hologram on the respective panel. For characterizing temporal or spatial fluctuations of the hologram on the panel leading to reduced stability of the optical traps, one analogue and one digital addressed spatial light modulator are compared regarding the aforementioned issue. In this regard, it is known that digital addressed SLMs usually show higher voltage deviations than the analogue devices typically leading to less temporal phase stability.

Phase Stability of SLMs and Related Effects

The capabilities of holography for precisely tailor the light of an optical trapping system in real time awoke significant interest especially since proposed in the late 1990s [83] that was mainly due to advances in semiconductor research and in digital technologies which became also more affordable. Holographic optical tweezers provide powerful means for dynamically generate complex 3D patterns of light foci and extraordinary beam shapes for a comprehensive versatility of manipulation experiments [95, 96].

The application of diffractive elements to divide a laser beam into subarrays of traps has been first implemented by Dufresne and Grier [97]. By placing a fixed, computer-generated hologram at the conjugate plane of an objective's entrance pupil, a predetermined pattern of traps has been created. Better concepts and systems of the basic idea quickly came along with the introduction of a spatial light modulator (SLM) into the beam path of a microscope setup. One of the most significant achievements was the updating of a phase kinoform on the liquid crystal (LC) display at video rate, that translated into real-time control of several traps independently [98]. Several methods to calculate the kinoforms efficiently at high rates or even video rates have been investigated. Consequently, a variety of algorithms that optimize many numerical procedures have been proposed [96, 99].

SLMs of parallel nematic liquid crystal type or phase only modulators offer an easy way to tailor light by allowing a precise modulation of phase of the incoming light without altering its amplitude [87] (see also Sect. 2.5). Data in form of two-dimensional grey-level images or matrices or masks are send to the panel via a controller, wherein the matrices typically includes significant contributions of the corresponding Fourier transform of the wanted pattern of traps [100]. Sources for an ultimate degradation of the trap quality especially in case of high dynamics are given by fluctuations of the parameters of the complex modulated optical wave [101, 102]. Therefore, the application of traps for quantitative measurements can be challenging in terms of reproducibility and repeatability [103] like it has been shown for force measurements by [104, 105]. Standard state of the art SLMs require the utilization of a varied voltage that can cause phase flicker (see also Sect. 2.5.2). One significant aspect is the refresh rate with which the kinoform is shown on the LC that is usually in the range of a video rate. For each new metrics or greyscale image the display is turned off for a short time [106, 107]. Another issue is the degrading performance of LCs caused by charge accumulation on the walls due to applied DC voltages[108]. Therefore, the control voltage requires a sudden flip usually in the range of a few hundred Hertz. The change in voltage can generate an unwanted modulation of the greyscale value that can be distinguished whether the LC is controlled analogue or digitally [109]. The molecules of the LC show a response only due to charge-balancing switching of an analogue switching device wherein continuous grey values are in principle provided. Contrary, in case of a digital control of the LC, solely discrete binary values can be adjusted which requires more sophisticated methods to change or refresh an exact control voltage at video rate. Especially the sudden switch between two states effectively leads to higher noise or fluctuations in each cell of the screen [109]. Typically, only two different voltages are available for digital backplanes, whereby each intermediate value has to be adjusted by pulse-width modulation (PWM), which means that the duration of a local square wave, that corresponds to the aforementioned on and off state, varies at a high-frequency in order to obtain the desired intermediate value in a temporal average.

Thus, referring to SLMs, a digital control leads to more fluctuations in the effective state of the liquid crystals directly affecting the precision of an optical tweezer. This drawback can partly be compensated by specially adapted control sequences, for example by applying smoothed changes in the voltage values or using the inertia characteristics of the liquid crystals. The experimental findings show that an analogue control of a SLM can achieve comparable results in terms of trap stability compared to non-holographic tweezers that often work, for example, with tiltable or movable lenses. The aforementioned increased stability of SLM driven optical tweezers opens up the potential to utilise holographic optical tweezers also for highly precise force measurements.

Experimental Setup

In order to investigate the temporal stability of laser traps modulated by spatial light modulators, two series of experiments are performed: first the stability of the phase shift or delay caused by the SLMs is determined. Afterwards, the impact of this phase delay on the trap position is determined by investigating the movements of holographically trapped particles.

Consequently, the findings from 3 different settings of the SLM are contrasted. The Holoeye modulator allows selecting several programmable addressing schemes that have a distinction in the number of available phase or grey levels and in its addressing speed. Thus, in addition to investigations with an analog Hamamatsu spatial light modulator (X10468 with 256 grey levels), two dissimilar digital addressing sequences are tested for a HoloEye SLM. The two aforementioned digital addressing sequences are named 5-5 (192 grey levels) and 0-6 (64 grey levels), respectively [110], whereby the first digit gives the number of equally weighted bitplanes, whereas the second digit gives the number of binary elements in the respective addressing sequence.

Measurements on Phase shift and stability The phase delay that is caused by an SLM can be determined with different techniques [111]. In this work, the straightforward technique is applied which is adequate for phase only SLMs [112]. In the setup that is used and shown in Fig. 4.31, a photodetector (photodiode) is positioned in the beam path behind a beam analyser A. The temporal phase changes can therefore be recorded by the photodetector with 15 kHz, especially when inserting two polarisers at $-45°$ and $45°$ before and after the SLM relative to the main alignment axis of the liquid crystals. The reflected light intensity I after passing the polarizer and the analyzer is proportional to the induced phase shift or delay ϕ of the modulator [26]:

$$I(\phi) = I_{\text{offset}} + I_0 \sin^2(\phi/2) \tag{4.45}$$

In this relation, I_{offset} and I_0 denote the offset intensity and the amplitude of the intensity modulations. The intensity fluctuation is recorded for different phase delays imprinted on the SLM that corresponds to different constant grey levels, whereby the average phase shift induced by a given grey level is related to the gamma curve. The measured intensity signal can then be utilized to calculate the phase as a function of time via (4.45). A helpful tool is given by the power spectra of the intensity.

Positional stability of beads With the help of polystyrene beads that are trapped with optical tweezers, fluctuations or possible instabilities of the optical traps can indirectly be measured and studied by recording the corresponding field of view with the beads with a high speed camera (PCO, 2500 frames/s). In typical studies, one or even a few beads with a diameter of 1–3 μm are trapped and observed throughout time. For the following investigations on the stability, a polystyrene bead of 2.5 μm in diameter is trapped on a positions outside the zero order and the motion of the bead is recorded. With automated image processing and analysis the center of each trapped bead and thus the deduced trap position can then easily be localized and tracked. During the investigations, an additional phase gradient with various spatial frequencies in x- and y-direction is generated for inducing lateral forces. Since the SLM models offer a higher resolution, a variation in x-direction is advantageous and also sufficient for the investigations due to an otherwise general symmetry of the optical path.

Experimental Findings

In order to assess and compare the temporal positional stability of holographic optical traps, the position of a digitally and analogue addressed holographic tweezer is tracked over time (see plots in Fig. 4.48(left column)). It can be deduced from the spots, that the digitally controlled SLM has higher positional noise compared to the analogue controlled SLM. In a first step the phase stability over time is investigated representing a key parameter for characterising the performance of optical traps that is directly correlated with the positional stability of a trapped object. Both SLM types show specific instabilities over time, wherein the digitally controlled SLM clearly results in significantly increased phase noise (Fig. 4.48, left column). The magnitude of the so-called flicker signal is correlated with the magnitude of the desired phase value, more specifically with maximised phase modulation amplitude revealed at distinct values for both types of SLM:

Digitally addressed SLM: (Holoeye)
For 0-6 mode: grey level=224, phase delay=1.75π;
for 5-5 mode: grey level=192, phase delay=1.7π
Analogue addressed SLM: (Hamamatsu)
grey level=64, phase delay=1.2π
Flickering
for the 5-5 0.31π
for the 0-6 settings 0.16π

The flicker's magnitude varies with the set phase delay, that is proportional to the grey level, with the phase modulations' highest amplitude that is present at a SLM specific phase value and setting (For Hamamatsu SLM: phase delay=1.22π, grey level=64; For Holoeye SLM: grey level=192, phase delay=1.71π for 5-5 mode; at grey level=224: phase delay=1.76π for 0-6 mode). The flickering can often be quantified to approximately 0.32π and 0.17π for the 5-5 respectively the 0-6 settings, and for the Hamamatsu to $510^{-4}\pi$. Accordingly, it is assumed that holographic traps generated with a digitally-addressed SLM will suffer from lower position stability due to the higher fluctuations. Nevertheless, a significant partition of phase amplitudes do not effectively influence the trap position (Fig. 4.49a). Contrary, the spatial stability is highly influenced by the modulations temporal phase at specific grey levels and/or the magnitude of the change of the amplitude (Fig. 4.49b).

Since the bigger amplitudes in modulation of the phase are of higher interest for a statistical analysis of the stability, in a next step the signal spectrum is calculated $I(\phi)$. This may allow deriving dependencies between undesired low stability - corresponding to the phase noise - and the frequency (Fig. 4.48, right column). It can be concluded that for the SLM with analogue addressing maxima occur at ≈ 240 and 480 Hz (Fig. 4.48a). These values correspond to the frequency harmonics of the addressing electronic board of Hamamatsu. The strength of this dependence can be further analysed by taking into account the spectrum of the laser alone. The laser's

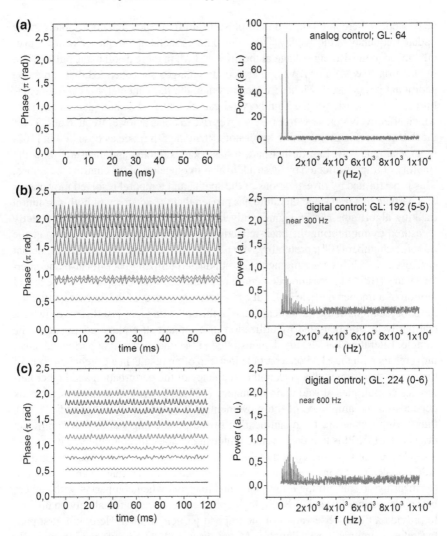

Fig. 4.48 Left column: for the analogue modulator phase versus time (**a**), for the digital modulator (**b**, 5-5 control mode; followed by 0-6 mode in figure part **c**). Measurements at different modulator grey levels: 0 (bottom), 32, 64, 96, 128, 160, 192, 224 and 255 (top). Right column: For every SLM control mode, a power spectrum of that data sample with the highest amplitude is depicted and illustrates a frequency influence of $I(\phi)$ modulation

characteristic flat spectrum is amended by high frequency detector noise at approximately 6 kHz. Further, a band pass filter is likely to be created due to a transparent semiconductor photo diode $f_{73\,\text{dB}} = 6.7\,\text{kHz}$ that is similar to the findings in [113].

Frequency maxima near the addressing frequency are also apparent for the digital SLM at $f_0 \approx= 300\,\text{Hz}$ for the 5-5 control mode and $f_0 \approx 600\,\text{Hz}$ for the 0-6 mode. Usually digital SLMs show significantly higher noise. Further, additional peaks appear, here at 60 Hz and multiples of said frequency that corresponds to the update frequency of the display. The occurrence of such high frequency peaks is usually linked to the driving voltage of the digital SLM, in more detail to the pulse-width modulation (PWM). For PWM, intermediate voltages are generated that can cause additional peaks also at higher frequencies than the fundamental, lowest frequency. Further, the viscosity of the liquid crystal causes lower noise at higher frequencies which effectively - on a relative basis - gives rise to noise at lower frequencies. Thus, the 0-6 addressing mode shows lower noise than the 5-5 mode as observed in [102] in an analogue way. In a further part, it has been examined and measured how the stability of traps is affected by these additional frequencies of the control electronics. This is performed by investigations of the spatial and temporal resolved movements of a particle in a trap since literature has proven that the accuracy of high-resolution tracking algorithms can be significantly limited when the trap position is directly measured by monitoring the back-reflection of laser foci from the coverslip of the trapping chamber [104], preferably with a high speed camera. In more detail, in that case usually stability measurements are conducted utilizing back-focal-plane interferometry (BFPI) for position detection, but which can fail comparing to the actual motion of a trapped particle [102] due to scattering effects of the particles, affecting the spot locations at the back focal plane of the collecting lens [104]. This drawback can partly be compensated by utilization of a pinhole at a plane conjugate to the sample. Literature has already demonstrated [114], that generating periodical small movements with the SLM on a particle in a trap often results in a Lorentzian form of the power spectrum with a peak at the frequency of the perturbation and higher harmonics at multiples of said main frequency [102], wherein the strength of the peak depends on the amplitude of the movement of the particle. (see Appendix 4.7.1), thus giving a measure for quantification of the positioning stability. A significant drawback of BFPI is that the measured intensity is normalized to the maximum and thus changes of intensity correspond to a change in position of the bead or particle. Additionally, little non-synchronous readout of deflection data and intensity can lead to smearing or convolution effects and thus to errors when comparing with camera based measuring. Instead of BFPI, high-speed cameras can alternatively be utilized to provide a direct observation of the trapped particle motion. Here, in a first part, the related findings for a digital SLM are discussed. The beads motion typically shows a spectrum depicted in Fig. 4.50 in dependence of the control modes, wherein the control or update frequency f_0 of the SLM causes high peaks at approximately 600 and 300 Hz mainly due to the digitally performed control. This is shown in the phase plots (Fig. 4.48), wherein the oscillation of the modulation can be expressed by: $x_{\text{trap}}(t) = x_0 \sin(2\pi f_0 t)$.

The power spectrum of the bead's motion may ideally be given by [114]:

$$P = \frac{D/\pi^2}{f^2 + f_c^2} + \frac{x_{\text{bead}}^2}{2} \delta(f - f_0) , \qquad (4.46)$$

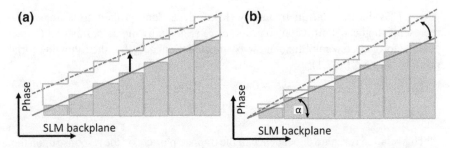

Fig. 4.49 a Modulation of the phase with the same amplitude that are all coherent (in phase) does not alter the phase gradient, generating the spatial frequency of a kinoform. Contrary, incoherent modulations (asynchronous) and/or **b** differ in amplitude can alter the phase gradient. In this schematic example, temporal variation of the angle α leads to a modulation of the trap position d, where f' is the objective's focal length and m is the magnification of the telescope used to image the SLM onto the objective's entrance pupil

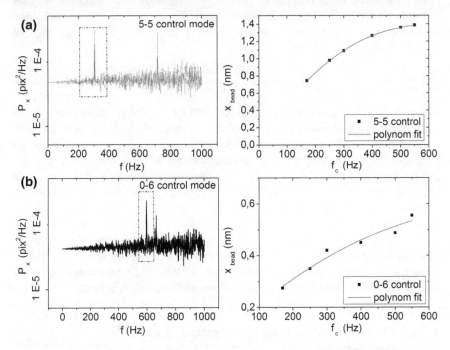

Fig. 4.50 Left column: power spectra of trapped bead position with variation in the SLM control scheme: **a** 5-5 and **b** 0-6. The peak of the SLM addressing scheme appears at $f_0 = 300\,\text{Hz}$ and $f_0 = 600\,\text{Hz}$, respectively (red arrows). The peak at approximately $720\,\text{Hz}$ is from the cooling fan of the camera. Right column: the bead's amplitude of the movement extracted from the maximum's magnitude versus corner frequency of the trap. A fit according to (4.47) delivers the trap movement x_0

where f_c is the trap roll-off frequency (linear dependent on the trap stiffness), D represents the bead's diffusion constant, $\delta(f)$ is given by the delta function (Dirac pulse). However, the amplitude of the bead oscillation x_{bead}, has the following form given by (see Sect. 4.7.1):

$$x_{bead} = \frac{x_0}{\sqrt{1 + \left(\frac{f_0}{f_c}\right)^2}} \tag{4.47}$$

This relation reveals a dependence of the trapped particle on the response upon the strength of the optical potential. Solely if the laser power is over a certain threshold for the viscous forces exerted by the surrounding fluid does the particle start to follow the trap. Further, then the phase fluctuations turn into position fluctuations. Also, the found closed mathematical dependence of the bead motion on the trap strength gives strong evidence that the reason underlying the appearance of noise in the power spectra is originating from trap motions, offering a further way for determining oscillation amplitude of the trap, x_0. The magnitude of the peak in each power spectrum at f_0 has been utilized to retrieve an experimental measurement of x_{bead} (Appendix 4.7.1):

$$x_{bead} = \sqrt{2(P_{Peak} - P_{therm} \Delta f}, \tag{4.48}$$

where P_{Peak} and P_{therm} are given from Appendix 4.7.1. Every single data point in Fig. 4.50a, b, right column, is calculated from one power spectral data, that is in turn the average of 50 measurements. The experimental findings were redone to get x_{bead} for traps with non equal corner frequencies, that were monitored with the laser power. The findings were fit by Eq. (4.47) to retrieve the value of x_0, as illustrated in Fig. 4.50. The value for the 5-5 setting, $x_0 = 1.66 \pm 0.25$ nm, is in accordance with previous findings [104]. Wherein for the 0-6 setting, a lower oscillation is shown, $x_0 = 0.84 \pm 0.14$ nm which is plausible related to the improvements seen in the phase measurements. Further, the good fit strengthens the physical model that the mechanism by that peaks show up in the spectra is indeed a rather physical oscillation of the trap. The error value for x_0 corresponds to the standard deviation of x_0 values determined in this manner from measurements on three different beads.

4.7.1 SLM-Calibration with Estimation of Particle Oscillation

Assuming no significant thermal noise, it is possible to retrieve the particle oscillation amplitude in (4.47) from the equation of motion ruling the dynamic behaviour of a system:

$$m\ddot{x}(t) + \gamma\dot{x}(t) + k\left(x(t) - x_{trap}(t)\right) = 0, \tag{4.49}$$

wherein the function $x(t)$ denotes the temporal position of the particle with mass m, the harmonic trap stiffness k, the viscous drag coefficient γ. Further, the trap motion can be given by a pure sine of the frequency f_S : $x_{trap} = x_0 \sin(2\pi f_S t)$. The viscos and thus damping property of the fluid usually leads to an overdamping of the inertial force of the surrounded particles [113], especially related to the typically applied time intervals of the measurements. This leads to a small Reynolds number ($Re \ll 1$) that confirms that the first part in (4.49) can be neglected. Accordingly, the final term can be reduced to:

$$\gamma \dot{x}(t) + kx(t)kx_{trap}(t) , \qquad (4.50)$$

The functional numerical standard solution of such a damped oscillation is usually given by a sine function of frequency f_s for the homogenous, stationary solution and an exponential decay function with a characteristic decay time $\tau \equiv 2\pi\gamma k$. The aforementioned parameter typically has values in the range of ms which means that it only needs to be considered at several kHz that is negligible for the time intervals considered in the utilized power spectra. This leads to the following solution of (4.50) consisting of the stationary term with an amplitude given by (4.47):

$$x(t) = x_{bead} \sin(2\pi f_S t + \phi) = \frac{x_0}{\sqrt{1 + \left(\frac{f_s}{f_c}\right)^2}} \sin\left(2\pi_0 t - \tan^{-1}\left(\frac{f_s}{f_c}\right)\right) . \qquad (4.51)$$

The parameter f_c, usually given by $f_c = k/2\pi\gamma (= 1/\tau)$, denotes the so-called roll-off frequency. Referring to the explanations in Sect. 4.7, the value of x_{bead} is effectively retrieved from the amplitude of the power spectral peaks. Consequently, (4.48) is utilized according to a qualitatively equal method of [114]. The temporal Fourier transform of (4.51) is given by:

$$\tilde{x}(f) = x_{bead} e^{i\phi} \frac{\delta(f + f_s) - \delta(f - f_s)}{2i} , \qquad (4.52)$$

The parameter $\delta(f)$ denotes the Dirac delta function (dirac pulse). Thus, the one-sided power spectrum part related to the oscillation equals to a delta pulse at the frequency f_s:

$$P(f) = \frac{2 |\tilde{x}(f)|^2}{T_{msr}} = x_{bead}^2/2 \, \delta(f - f_s) .| \qquad (4.53)$$

The time for the measurement of each power spectrum is denoted with T_{msr}. The surface under the function can be expressed by $x_{bead}^2/2$, which is related to the measurement retrieved from the area under the peak, A, according to:

$$A = \left(P_{peak} - P_{therm}\right) \Delta f = \frac{x_{bead}^2}{2} . \qquad (4.54)$$

First, the background P_{therm} of the thermal noise is defined by the integral of a Lorentz-fit function applied to the power spectrum. Then, the peak area P_{peak} is integrated above the aforementioned calculated background and the frequency spacing is given by $\Delta f = 1/T_{\mathrm{msr}}$. With the help of that equation the desired value of (4.48) can be calculated.

4.8   Integration in Nano-positioning System

A variety of microscopy techniques allow measuring different local physical properties of a surface under test. One of the key properties of interest in production and development of micro- and nano components is a nanometre resolution even in a measurement range of a few centimetres. By integrating a low coherent digital holographic microscope (DHM) into a coordinate measuring machine with sub nanometre resolution and nanometre uncertainty, a DHM with an outstanding measuring range is realised which enables simultaneous investigation of form and roughness of specimens with sizes up to $25 \times 25 \times 5\,\mathrm{mm}^3$ along the x, y and z-axes (see also Sect. 3.6).

Different modes of scanning strategies have been analysed and error compensated for micro and nano structured optical components with a surface diameter up to 25 mm.

There are growing demands to test the dimensions of micro optical components and on tiny features that are constituencies of more conventionally sized specimens. The type of object can vary significantly, from micro-fluidic channels, micro-lenses and fuel injector nozzles to features and structures on micro-electromechanical systems (MEMS). The increasing demands made by microelectronics, optics and precision engineering industries to reduce the measurement uncertainty in the nanometre range with the volumes getting larger and larger are the key motivation for the development of different concepts of nano-positioning and nano-measuring systems. Thus there are still new systems being developed for increased large-area scanning probe microscopy or for their use in the geometric characterisation of meso-scale objects with nanometre uncertainty [115, 116]. Till now, various tactile, minimal invasive and non tactile probes based on different measurement principles have been integrated in nano-coordinate measuring machines (nano-CMMs) [117]. This comprises profilometers, white light interferometry, focus sensors, confocal microscopes and atomic force microscopy as well as other related SPM-techniques.

Significant measurement problems often arise because the coordinate measuring technique does not fit to the properties of the surface of the specimen. In many cases, the probing system is the limiting factor of the measurement system: Either it is not possible to access the feature which represents the main challenge for optical and SPM systems, or the forces associated with tactile CMM-like probes damage the surface or feature. In practice, however, there are many different requirements for sensing surfaces or for detecting structures. There are several physical principles according to which probes can be realised. Despite there is a wide range of probes utilised for nano-scale measuring systems, these are not generally appropriate for measuring three-dimensional (3D) features. In many cases, the probing systems intended for measurements of micro-features are miniaturisations of rather conventional microscopy techniques or CMM probes being adopted for 3D capability and better repeatability, which opens up the potential of precise calibration. Thereby, the features of DHM methods (see Sect. 4.2) enable a subsequent focus and thus a focusing of different sub-areas of the measured field which offers the possibility of faster measurements in a stitching mode over the whole positioning volume of the nano-CMM.

4.8.1 Experimental Investigations on Sensor Integration

The adjustment of the effective coherence length shown in Sect. 4.3.5 is now applied for adapting the fringe number available for temporal or spatial phase shifting digital holographic microscopy. This allows - depending on the measurement task - a more or less sharp selection of the measurement layer when testing e.g. stacks of different optics and an elimination of multiple reflections when automatically measuring extended surfaces in a nano-CMM in a stitching mode. The application dynamic adjustment of the available fringe number is demonstrated on a micro lens array (Fig. 4.51) with different lens sizes (smallest lenses: Pitch: $114.3\,\mu m$, $NA = 0.1$, $PV = 2.83\,\mu m$).

While e.g. the smallest lenses only require a coherence length of about $3\,\mu m$, the medium sized micro lenses need a coherence length with a factor of ≈ 2. When testing those micro lens arrays, this results in higher noise values, calculated according to [36] (see Fig. 4.51). Thus, the optimal noise level can be adjusted dynamically. Reference measurements have been carried out with an atomic force microscope to verify that both sizes of the lenses have equal roughness values of $(4.5 \pm 0.8)\,nm$ to exclude systematic errors.

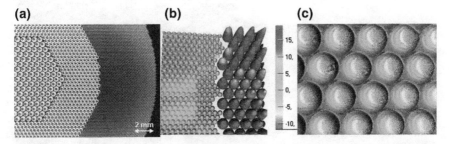

Fig. 4.51 **a** Microscopy-image of microlens array where the smallest lenses have a pitch of $114.3\,\mu m$ ($NA = 0.1$, $PV = 2.83\,\mu m$). **b** Transition area from smallest to medium sized micro lens arrays, measured with adjusted coherence length for temporal phase shifting digital holographic microscopy (noise level: 8 nm). **c** Biggest lenses measured with strong mismatch of spectral widths in the interferometer which causes a noise level of approx. 32 nm

4.9 Realisation and Illustration of Beam Configurations

As already described in the theory Sect. 2.6, according to the principle of holography arbitrary wave fronts may generally be produced by a corresponding phase mask on the SLM. The visualisation of the beam configuration can be achieved with a detector screen. Since the dichroic mirror reflects the infrared light, which subsequently can not reach the camera, a fluorescent detector card may be inserted into the microscope object plane. The excited fluorescent material of the card emits a different, shorter wavelength which then can pass the dichroic mirror.

To record the intensity distributions of the individual rays in a high effective resolution, an additional setup is devised similar to Fig. 4.41. This consists of a fibre-coupled 532 nm laser, a lens (L1, $f = 200$ mm) for the collimation of the beam, an adjustable polarisation filter, a neutral density (ND) filter, a variable iris, two additional lenses (L2 with $f = 160$ mm and L3 with $f = 60$ mm), a CCD camera and the SLM (see Fig. 4.52). The deployed laser has a wavelength in the visual area of the electromagnetical spectrum and therefore the gamma curve of the light modulator has to be specifically calibrated for that wavelength since the induced phase shift is a function of the optical path length difference $\Delta\phi = \frac{\Delta s}{\lambda}2\pi$. The gamma curve determines the needed voltage for the individual pixels of the SLM which corresponds to the required phase shift (see also Sect. 4.6).

In combination with the filters and the iris, the first lens expands the beam and homogenises it. The incident wave on the SLM is therefore nearly plane. The ND and polarisation filter ensure that the camera is not driven into saturation and the iris diaphragm controls the area of the SLM that is utilised to prevent unwanted and obstructive reflections.

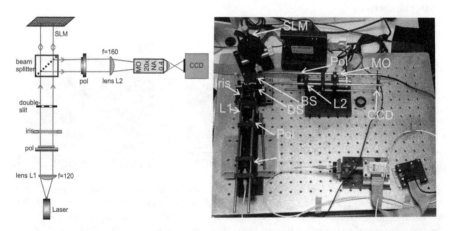

Fig. 4.52 Setup for the acquisition of the intensity distribution of the different beam configurations

4.9.1   Bessel Beams

The realisation of a simple Bessel beam is possible in two different ways: First, this beam can be generated with a circular hologram on the SLM, developing an OAM Bessel beam (with an angular momentum: Sect. 2.9.2). Alternatively an axicon can be used as hologram.

In order to generate superimposed or higher order Bessel beams resulting in an OAM-beam which is configurable through different parameters like the radii of the rings and the azimuthal order (see Sect. 2.9.2), the setup depicted in Fig. 4.52 is utilised. Similar to Fig. 2.23, the relation between the phase mask on the SLM, the intermediate plane (IF), the far field plane (FF) and the Fourier or near field (NF) is illustrated in Fig. 4.53. The simulation includes multiple 2D and 3D images of the intensity distributions. The number of maxima and therefore the azimuthal order can easily be counted in 2D as well as in 3D representation.

The experimental proof of the intensity distribution in the transition area corresponding to a specific ring hologram is depicted in Fig. 4.54. An analogue intensity distribution to the phase mask is visible in the recorded intensity distribution. Furthermore the spiral nature of the wave, the vortex or the discrete transition from 0 to 2π can be identified. Despite the fact that the SLM was only assigned a circular pattern (with a reflection of 0 within and outside of the ring), some speckles and noise signals appear within, even when the correction of the SLM topography is used (see Sect. 4.6). This is caused by a basic reflection of the light modulator. To counteract this effect, a diffuse "disruption" pattern can be generated as background [118, 119]. This diffuse disruption pattern, which will be referred to as "checkerboard", suppresses the parasitic signal and leads to a significantly higher quality of the wave fronts and thus intensity patterns. An additional recording with and without suppression of the reflection utilising a camera with a broader dynamic range in Fig. 4.55 shows the similarity to the simulated results of the intermediate plane. A simple axicon displayed on the SLM - as a correspondence to the refractive optical element - described by Eq. 2.98 is investigated as well. With this method the whole area of the SLM is utilised and therefore no unwanted or random reflections occur. Thus the generation of the "checkerboard" is not applicable. Furthermore the main maximum of the intensity, the characteristic feature of the basic Bessel beam, is relatively distinctive and can be identified in Fig. 4.56(right).

(a) <u>Ring holograms for generation of Bessel beams</u>

single ring   double ring
  l=3   l=1   l=2   l=3   l=4   l=5

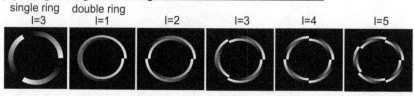

(b) <u>Simulation of intensity distribution for different z-planes 2D</u>

single ring   double ring
  l=3   l=1   l=2   l=3   l=4   l=5

near field:

intermediate plane:

far field:

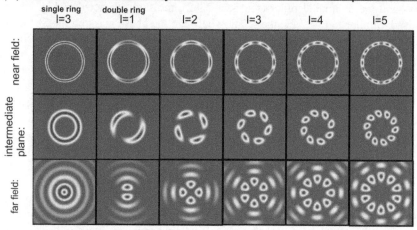

(c) <u>3D representation of fields shown in (b) for double rings:</u>

double rings
  l=1   l=2   l=3   l=4   l=5

near field:

intermediate plane:

far field:

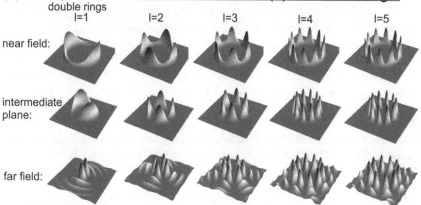

Fig. 4.53 Simulated normed intensity distributions of selected Bessel beams for the different z planes (near, far and transition plane) in **b**, **c** with respect to the azimuthal parameter l with the corresponding phase masks modulo 2π in **a**

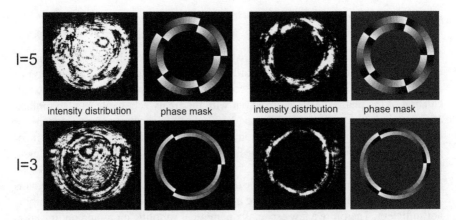

Fig. 4.54 Experimental Bessel beam intensity distribution and the corresponding phase mask modulo 2π in the transition plane with $l = 3$ (below) and $l = 5$ (on top) without (left) and with (right) checkerboard phase patterns in desired inactive regions of the SLM panel

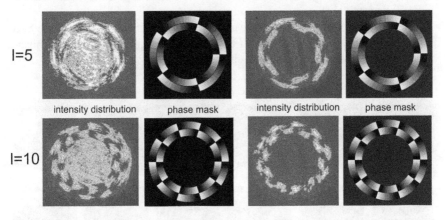

Fig. 4.55 Experimental Bessel beam intensity distribution and the corresponding phase mask modulo 2π in the transition plane with $l = 10$ (below) and $1 = 5$ (on top) without (left) and with (right) applied checkerboard

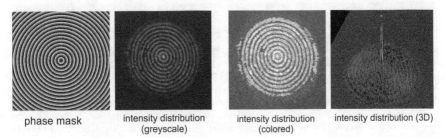

Fig. 4.56 Bessel intensity distribution with corresponding phase mask (left) in a monochrome, coloured and 3D representation (right), where the main maximum is clearly visible

4.9.2 Mathieu Beams

The generation of the Mathieu beams is investigated in the following. Therfore, an algorithm has been implemented according to Sect. 2.9.5 allowing the variation of order and parity for the display of the corresponding phase onto the SLM. The calculation of the phase mask is carried out with the freely accessible Matlab toolbox of Cojocaru (see equations (2.128) and (2.129), [120]). The computer operations for this beam type require high calculation resources. The mask consists of two independent matrices for the radial and the azimuthal part. For the radial part the Bessel function of a matrix with a resolution of 1920 × 1080 hast to be calculated. This can take up to one minute dependent on the order of the configuration.[1] Furthermore this operation can not be performed above a certain order since the memory capacity is not sufficient. Therefore a few especially numerical simplifications and optimisation steps have been introduced.

The theoretical phase pattern (hologram) and theoretical as well as experimental intensity distribution of different selected Mathieu beams is given in Figs. 4.57, 4.58 and 4.59 respectively.

The phase mask of the Mathieu beam also utilizes a closed surface of the SLM area to full extent and is - in contrast to different configurations - "quasi"-binary, since the theoretical phase values are either 0 or 2π.

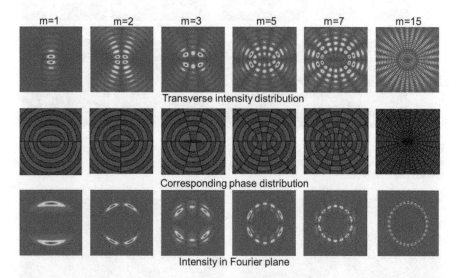

Fig. 4.57 Mathieu beam patterns with theoretically calculated intensity distributions and corresponding phase masks for different orders with even and uneven parity. First line: Transverse intensity distribution. Second line: Corresponding phase distribution. Third line: Intensity in Fourier plane

---

[1] System configuration: Intel Pentium Core Duo 2.13 GHz, 3 Gb RAM, GeForce 210 with 1024 Mb graphic memory.

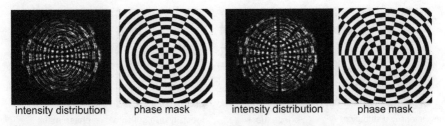

intensity distribution   phase mask   intensity distribution   phase mask

Fig. 4.58 Mathieu beam with experimentally determined intensity distribution and the corresponding binary phase masks $[-\pi, \pi]$ for the 5th order with even parity (left) and 8th order with uneven parity (right)

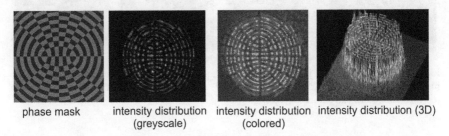

phase mask   intensity distribution   intensity distribution   intensity distribution (3D)
                  (greyscale)   (colored)

Fig. 4.59 Mathieu beams of 8th order with the corresponding phase mask that is corrected by the "checkerboard" (left) and the corresponding normed intensity distributions in different representations

By calculation and application of the diffuse disruption pattern according to the "checkerboard" [118, 119] principle in outer regions of the SLM, also here the intensity distributions appear clearer and the spurious reflections are being reduced.

Because of their symmetric distribution of sharp spots, the Mathieu beams allow a light efficient stable trapping pattern without ghost traps which would occur for the conventional generation of a similar pattern according to the lenses and gratings algorithm discussed before.

4.9.3 Laguerre Beams

Laguerre beams are beam configurations in this section that are explicitly denoted as non-diffractive and still transfers an angular momentum. They are generated with two parameters that can be chosen arbitrarily: The radial (number of 2π) and the azimuthal index which are previously defined and then transmitted to the SLM. The calculation of the phase mask is mainly done by the native Matlab Laguerre functions. Similar to the calculation of the Mathieu beams a resource problem with the PC's hardware is possible at this point since the implemented numeric Laguerre function realised in Matlab can not be further optimised with regard to calculation time and thus is very time consuming for the high resolution of the applied SLMs.

The intensity distribution of the Laguerre beams likewise show spurious reflections in Fig. 4.60.

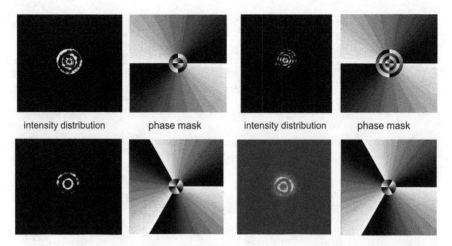

Fig. 4.60 Normed Laguerre intensity distribution and the corresponding phase masks modulo 2π for $l = p = 2$ (top left) $l = 2$ and $p = 3$ (top right) and $l = p = 2$ (bottom) in a monochromatic representation (left) and coloured (right)

Nonetheless a typical Laguerre beam is recognizable. The similarity to the Bessel beam of first order that was generated by an axicon (see Fig. 4.60(bottom line)) is especially distinctive. The order of the Laguerre beam can be determined by the number of the rings. The angular momentum (spiral feature) is identifiable in the second line of the images in Fig. 4.60.

In general it is shown that the Laguerre beams especially of lower orders may also be applied as a fixed trapping pattern, but with optical torque which may serve in special micro-fluidics based investigations in biomedics.

4.9.4 Airy Beams

To generate the Airy beam no highly relevant parameters are required in comparison to the precedent beams. The phase mask of the airy beams generates the intensity patterns depicted in Fig. 4.61. Since they cannot be customised and changed significantly, this beam configuration is solely of little importance, but can be valuable for exerting a high force gradient e.g. for dissolving agglomerations of cells or micro beads.

Analogue to the latter beam types, the Airy beam is demonstrated successfully with the system. Since every beam type has its individual force and torque pattern, these can now be selected for some adapted applications which is performed in the following chapter.

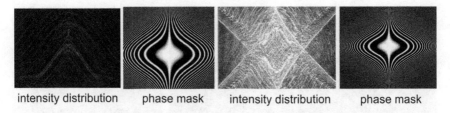

intensity distribution   phase mask   intensity distribution   phase mask

Fig. 4.61 Normed Airy intensity distributions and the corresponding phase masks modulo 2π

4.10 Application of Trapping Patterns and Optical Torque

In general, any beam with inclined wave fronts carries orbital angular momentum, e.g. Laguerre–Gaussian laser beams, which have a helical phase structure described by $\phi = l\theta$ (see also Sect. 2.9.3). At the centre of the helical phase is a singularity where the phase is undefined and the field amplitude vanishes, giving rise to a "dark beam" in the core of the light wave. The orbital angular momentum in a helical beam (also called an optical vortex beam) can be adjusted by either changing the wave fronts' helicity l or by increasing the photon flux. Thereby, in [121] it has been demonstrated that the interference of two Laguerre–Gaussian beams of opposite handedness create interference patterns for micromanipulation of particles and holographic optical tweezers. Contrary, the Airy beam spot is built up of a bright main spot and several side lobes whose intensity increases towards the main spot (Fig. 4.62). Therefore, micro-particles and cells experience a gradient force which drags them into the main spot. Due to the light pressure exerted, micro-particles and cells are then levitated and propelled along the curved trajectory of the main spot away from the cleared region.

Even though they have the interesting property of propagating without apparent spreading due to diffraction, Bessel beams are non-physical, because, as uniform planes waves, they carry infinite power. This is a consequence of the fact that Bessel functions are not square integrable. Bessel-Gauss (BG) beams are the finite-power, physically realizable version of Bessel beams. As they can propagate over a certain distance without significant spreading, BG beams are often said to be nearly non-diffracting. Higher order Bessel beams may be superimposed in such a way that they produce a field which either has or does not have a global optical momentum [122]. When generating a superimposed higher-order Bessel beam, which results in no global angular momentum, a rotation in the field's intensity profile as it propagates has been demonstrated. The superpositions of light fields are encoded on a spatial light modulator (SLM) with one ring-slit hologram (which may also be generated by illumination of an axicon with an overlay of Laguerre–Gaussian beams).

In the following, demonstrations of the near, intermediate and far field intensity profiles (see also Sect. 2.9.3) of a ring-slit aperture generated with a liquid crystal on silicon spatial light modulator (LCOS) in a digital holographic microscope are shown. For optimal harnessing of these special beams for holographic optical traps,

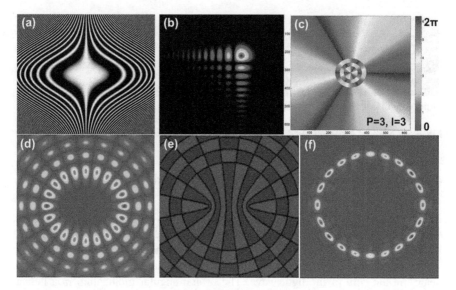

Fig. 4.62 a Airy beam phase distribution on SLM and corresponding intensity distribution in Fourier plane (**b**). Laguerre-beam phase pattern in **c**. Even Mathieu beam of order $m = 10$ and corresponding transverse intensity distribution of the non-diffracting beam in **d** (normalised and in false colour). Corresponding phase distribution in **e** (bright and dark grey values correspond to a phase of 0 and π). Intensity in a Fourier plane (**f**)

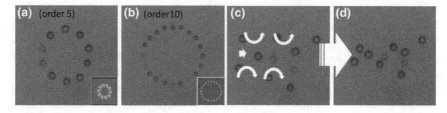

Fig. 4.63 a Superimposed Bessel beam of order 5 for trapping of micro beads (1 μm diameter) in the far field (FF). **b** Even Mathieu beam of order $m = 10$ to trap 20 particles in an elliptic circle. **c–d**: demonstration of generation of a particle flow: changes within 10 s

knowledge of the structure or intensity profile of the complex beam at various planes in the area of the sharp image plane is required.

In order to demonstrate the applicability of the non-diffractive beams, the intensity patterns of superimposed Bessel beams and Matthieu beams are exemplarily shown for micro beads, but may be applied to hold biologic cells in fixed equally distanced positions for time lapse experiments where apoptosis is induced and forwarded by messengers (Fig. 4.63a, b). Furthermore, the multiplexed superimposed Bessel beams are applied to generate a particle flow (Fig. 4.63c, d). The different applications differ solely in the axial distance of the field which is applied. Thus in order to evoke optical torque, only the distance has to be changed slightly. On the other hand, this torque

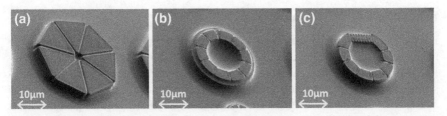

Fig. 4.64 SEM measurements of different geometries of micro-fluidic structures produced by 2PP direct laser writing adapted to the order of specific Bessel beams in order to allow higher effective forces for fixing biologic cells on predefined positions indicated by red circles

has to be carefully considered in sensitive applications when approaching the Bessel beam to a tiny object. The use of higher order Bessel beams is also demonstrated in Fig. 4.64. Here, the geometry of micro-fluidic structures is adapted to the order of a specific Bessel beam in order to allow higher effective forces for fixing the cells on predefined positions.

In a further step, the proposed approach for real-time, non-invasive identification of micro/nano organisms and cells is demonstrated by integrating the two methods of digital holographic microscopy and holographic optical tweezers, which can be combined to a single compact hardware, to simultaneously sense, identify, control, and track cells and micro/nano organisms in three dimensions. New possibilities that arise from the proposed new method are discussed especially in Sect. 4.11.

4.11 HOT-DHM-Combination

In this section, first the two previously demonstrated optical techniques digital holographic microscopy (DHM) and holographic optical tweezers (HOT) are first briefly summarised with regard to several applications. The two optical methods enable minimal-invasive imaging, manipulation and identification of preferably micro sized biologic specimen in three dimensions. Since e.g. biologic cells solely change their properties slightly in dependence of health and treatment with drugs, here innovative approaches are suggested for identifying micro or nano organisms and cells by combining the two measurement methods of DHM and HOTs. Both methods can operate simultaneously in one common optical microscope path. Thus the envisaged system is capable of sensing, controlling, identifying and tracking biologic cells or other micro-organisms in 3D at the same time. This opens up prospect for innovative methods that are demonstrated and evaluated such as new methods for cell assembling especially in micro-fluidic devices and cell characterisation in the (sub) micrometre range. In particular, statistical algorithms implemented for tracking micro-organisms in 3D [123] and recognition of different characteristics of cells using DHM [124] have been tested. The so called *escape force measurement* using optical tweezers is introduced here as well for classification between different types of cells [125].

Brief Overview of DHM and HOTs

For the manipulation, identification and characterisation of biological specimen like micro/nano-organisms and cells there is an increasing interest in low-cost minimal-invasive and high-throughput methods. These methods are often based on a microscope and can have a significant role in key areas like next generation point-of-care health solutions, cancer diagnosis, environmental monitoring, food safety and fast detection of pandemics. Usually biological micro/nano-organisms show a complex system behaviour and interaction with their environment despite their often simple and minute morphology. In most cases, applications in biological characterisation are dominated by bio-chemical processes like fluorescence, which are often invasive and time consuming. Thereby, optical methods provide a balanced selection of sensitivity, non-invasiveness, low cost, compactness as well as speed. Thus, in the precedent section the combination of two optical methods based on holographic principles [126] for marker-free, real-time trapping, identification and tracking of micro-organisms is treated.

Typically, in classic microscopy semi-transparent cells need to be labelled with fluorescent markers or stained for imaging and characterisations in highest optical resolutions. Often, these invasive imaging techniques are frequently harmful for the cells or adversely influence their viability and disturb the cells' natural life cycle. As a result, this makes further assessments difficult in certain biochemical applications such as stem cell screening. To remedy some of these limitations, three-dimensional (3D) microscopy techniques have been investigated [127]. In particular, recent technological advances have permitted increased use of coherent optical systems [128, 129] mainly due to availability of a wide range of inexpensive laser sources, advanced detector arrays as well as spatial light modulators. An innovative approach for real-time, minimal-invasive identification of micro/nano-organisms and cells is supposed by combining the two methods of DHM and HOTs that is demonstrated in the following.

As shown in the previous chapters, optical microscopy methods can offer flexible tools for moving, rotating, trapping or sorting micro- or nano-sized objects with a minimally invasive approach that excludes mechanical contact. The corresponding forces are generated by a gradient vector field of optical forces or the optical momentum an object is exposed to if it deflects a flux of photons [130] (see also Sect. 2.6).

As mentioned in Sect. 2.6.1, in contrast to the double beam optical traps, those with a single beam are relatively easy to realise and integrate in available microscope setups. Therefore, a single microscope objective with a high NA is required that is usually fulfilled for oil immersion objectives with a magnification factor of $>40\times$. The drawback of this trapping geometry is the relatively high intensity required for successful moving of objects which could harm biologic cells. Additionally, the required high NA objectives have only a small working distance and low depth of field of a few micrometers where the intensity is high enough.

In order to reduce a possible damage on objects and for minimisation of stray light, usually NIR-light in the range of approximately 780–1330 nm is utilised.

Furthermore, in this wavelength region the absorption is small in comparison to lower wavelengths due to the cell's components, but also at the higher wavelengths due to water absorption [72].

According to [131], light at 830 and 960 nm is particularly adequate, but due to lower costs of widely available lasers with a wavelength near 1060 nm [132]. Often the specimen is varied in position with a microscope xyz-stage and the focal spot stays static in a typical single trap system. Alternatives are given by steering mirrors [133] or acousto-optical beam deflectors [134] that allow the creation of several independent traps based on time-sharing or beam multiplexing (Sect. 3.1).

In contrast to the latter approaches, holographic optical tweezers (HOT) utilize a spatial light modulator (SLM) to create multiple traps that can be used independently. HOTs even allow to produce special intensity distributions with solely little restrictions (see also Sect. 4.10). As demonstrated in the previous chapters, this approach offers the possibility of various trapping configurations and 3D geometries as well as aberration control.

DHM-HOT Based Cell Characterisation

In this section, a summarizing overview of applications of DHM-HOT approaches for characterisation and tracking preferably of biologic cells or small organisms is given. For tracking biologic specimen in three dimensions, statistical algorithms have been investigated [123] and identification based on image pattern recognition of various species utilizing DHM have been explored [124]. The so called *escape force measurement* applying optical tweezers as explained in the next section has also been reported for a distinction between non-cancerous and cancerous cells [125]. Similar to biometrics, biologic cells typically show special characteristic that can be used for identification. Here, for example the development of certain characteristics regarding their size, 3D shape, survival and response to environmental conditions is of typical interest. Moreover, the dynamic behaviour of multiple parameters for determining growth and possible cell division or a further cell development is of special importance which typically requires time lapse experiments. A subsequent data analysis of an acquired sequence of measurements may then allow quantitative results. This approach requires comprehensive data like multi content imaging techniques to develop sufficiently selective parameters which often cannot be provided by conventional 2D imaging techniques in contrast to DHM that allows an automation of selection processes [124, 135, 136].

In digital holographic phase contrast, a 3D representation of the phase change induced by a specimen is useful to determine the local refractive index of microorganisms and therefore to study the exact interaction of light and matter [137] as shown also in Fig. 4.65.

The retrieved DHM measurements are valuable for a variety of selection processes or diagnostics. Therefore, several statistical approaches have been investigated to segment and track especially biologic cells when they are imaged by a digital holographic system [123]. Here, e.g. a bivariate Gaussian statistical distribution is applicable to efficiently simulate the complex wave front information in a reconstructed electromagnetic field in the neighbourhood of the specimen. A maximum a posterior (MAP)

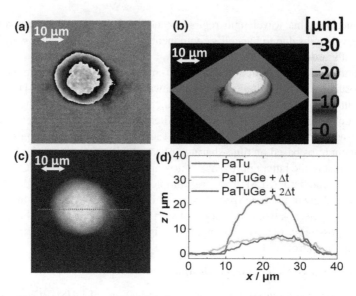

Fig. 4.65 a Phase distribution modulo 2π of a Pancreas tumour cell PaTu8988Twt before treatment with Gemcitabine. **c** Unwrapped phase distribution and corresponding pseudo 3D representation in **b**. **d** Profiles of the cell height through the centre at two different time intervals $\Delta t = 1h$

Bayesian framework has been applied in [123] for guessing the optimum refractive index distribution for a quantitative pseudo 3D ($2\frac{1}{2}$-dimensional) reconstruction of a (semi-)transparent specimen.

A DHM reconstruction of the specimen in a sequence of images at different z-positions also allows for an exact tracking of the specimen. This sequence can then be processed in time by the MAP most likely estimator for tracking the position and 3D representation of a biologic organism in a current hologram frame. In general it has to be considered that the shape of e.g. biologic cells is relatively sensitive towards the environmental conditions and may exhibit high statistical fluctuations.

The MAP estimator locates the optimal key parameter for the 3D representation of a cell and it's boundaries that enclose it on the basis of the DHM reconstruction. In case of the supposition that the cell boundaries can be described with a binary window, the corresponding force function changing the position of the specimen at time t, $\tilde{\varphi}(\vec{r}, t)$, is

$$\left|\vec{F}(r)\right| = \sum a \in \eta \omega(a) \log D_{\text{ta}}(a) - \sum_{a \in \eta}(1 - \omega(a)) \log D_{\text{ba}}(a) + \nabla \cdot (\vec{\nabla}\tilde{\varphi}(r, t)) \,.$$

(4.55)

Here, the parameter η denotes a spherical neighbourhood around $r = (x, y, z)$. Also, D_{ta} and D_{ba} denote bivariate Gaussian distributions for target and background, respectively. The last term in (4.55) embeds the prior information, which in this case is a term that penalizes sharp corners [123]. At every step, the level set is updated according to $\tilde{\varphi}(r, t + 1) = \tilde{\varphi}(r, t) + F(r)\vec{\nabla}_{ijk}\tilde{\varphi}$, where $\vec{\nabla}_{ijk}$ denotes the gradient operator.

Figure 4.65a, b depicts a phase hologram with an extracted profile in (d) marked with a white line in (b). The retrieved information concerning the 3D position and shape of the cell is helpful to generate optical traps at a correct position to manipulate e.g. biologic cells. A frequently required task for automated microscope screening systems is the characterisation of micro-organisms. In the following section, two approaches are treated. First, classification of cancerous versus non-cancerous cells utilizing the previously applied escape force measurement method with HOT is reviewed [125]. Second, a statistical approach is used to DHM reconstruction images that is based on a non-parametric model for enhanced discrimination between different micro-organisms [124].

Discrimination by Trapping Forces

The influence and interaction of optical traps on cells also depends on the environmental chemical conditions, structure and refractive index of cells as well as the surrounding medium. Usually, diverse cells, or similar cells in non-equal states, show different characteristics and therefore varying interaction or response in case they are exposed to an optical trap. One possibility to harness this effect for cell classification is the measurement of the three dimensional cell elasticity or deformation in an optical cell stretcher [138]. A selection and optical discrimination determined by the dynamics of the cells in a focal laser light spot has been demonstrated by [139]. The applied method for cell distinction here makes use of the change of the trapping force among a variety of biologic cells [125].

The utilised escape force method based on the measurement of the maximum achivable speed of a trapped particle in a fluid and according to [134] allows a precise force measurement by moving the position of the trap with increased velocity until the cell leaves the trap caused by the rising flow drag. The optical tweezer force is measured indirectly by determining the highest cell velocity during that half automated process. The 3D coordinates as well as the diameter d of the specimens can be retrieved via signal processing from a camera. Subsequently - by temporal differentiation of the cell's position - its maximum velocity v_{max} can easily be calculated. Applying Stokes law and the dynamic viscosity η_v of the medium, the trapping force can finally be retrieved

$$F = 12 \cdot \pi \cdot d \cdot \eta_v \cdot v_{max} \qquad (4.56)$$

Here, Stokes law is only applicable because of the small diameter of the specimens of $d \approx< 12\,\mu\text{m}$ and the approximately spherical shape. In case the cells are not spherical, the trapping force can be significantly different (see also Sect. 2.6.2). Then the force can be calculated based on the pseudo 3D representation determined by the DHM phase contrast method. In order to proof the selection functionality of the proposed method, these force measurements are explored on pancreas tumour cells (PaTu8988Twt) in three different cell states: Normal pancreas cells (*NoCe*; normal cells), their transformed counterparts (*PaTu*) and Gemcitabine treated cells (*PaTuGe*). Gemcitabine denotes a drug for cancer treatment [140].

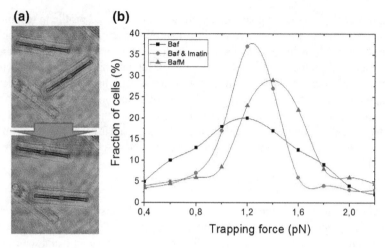

Fig. 4.66 Trapping force distributions of "normal cells" (NoCe), "cancer cells" (PaTu) and Gemcitabine treated cancer cells (PaTuGe, 4 h after drug exposure)

For the experiment, the cells are located between two cover glasses separated by a spacer at around 60–70 μm above the glass surface. Typically biologic cell concentrations of ≈ 100 000 cells/ml are taken for the measurement series. A quantity of 300 single cells of each cell state with sufficient distance to neighbouring cells in order to avoid influences has been trapped Fig. 4.66. The Gemcitabine treated cancerous cells are measured 4 h after drug exposure. A Gaussian function can then be subsequently fitted on the trapping force distribution. The optical trap force distribution of normal cells and cancerous cells can be separated by the individual standard deviation since the cancerous cells exhibit a significant broader standard deviation: 2.9^{-10} N in comparison to normal cells 1.7^{-10} N. Thereby, the histograms of drug treated PaTu cells and untreated PaTu cells show different locations of the peak. Throughout the measurement series, cancer cells led to a lower peak value than drug treated cancer cells.

A subsequent comparison with a modern commercially available fluorescent label based flow cytometer (FACS), the performed trap and escape based approach demonstrates a higher sensitivity for a distinction of Gemcitabine treated and untreated *PaTu* cells.

Figure 4.67 also demonstrates, that - after different incubation times in flow cytometry measurements of the Gemcitabine treated PaTu cells - it is not possible to distinguish untreated and Gemcitabine treated cancer cells after 4 h of incubation. The exact reason for the variation of the trapping force in different cell states cannot be given yet in detail, but it is mainly correlated with the shape and size of the specimen. Therefore, this may mainly affect the momentum transfer and thus the optical force on the biologic cells. A possible change of the refractive index might also affect the optical force. According to investigations on the integral refractive index of the cell types performed with a method analogue to [141], the cancerous cells

Fig. 4.67 Fluorescent marker based flow cytometry measurements of untreated (PaTu) and Gemcitabine treated cells (PaTuGe). TMRM and Annexin V are markers for early cell death processes

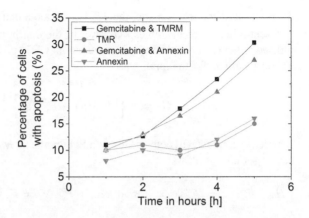

showed a slightly higher refractive index with $n_{\bar{P}aTu} = 1.375 \pm 0.004$ in comparison to $n_{NoCe} = 1.393 \pm 0.004$. Another factor might be a difference in the complex nanoscopic absorption distribution determined by the ensemble of all cell components like mitochondria et ceterae. Besides the proven applicability of the trap force distinction method, another advantage of this method is the relatively easy automation of the measurement process mainly based on image processing and an additional microscope stage control. Multiple traps have also a significant potential to speed up the measurements by a multiplexing of the whole process. Since the applied technique is marker free, no extended time and costly preparation of the specimens is required. Furthermore, an influence by markes on the result can be excluded here, which are often toxic.

Cell Identification Using Statistical Techniques Based on DHM

For the discrimination of similar cells, in addition to trapping forces as previously shown, statistical tests are a common method that can be applied to the holographic reconstruction data in order to realise recognition of biologic cells against a known database of species [124, 135, 136]. Statistical sampling theory is performed for determination if the variation between two recorded data sets is significant or not. For statistical decisions, a hypothesis test can be conducted by constructing a corresponding statistical sampling distribution of a specific given test statistic. This can be given by a two category classification like a cancerous versus benign, pathogenic versus non-pathogenic, terminally differentiated versus pluripotency cells or in general for identifying two species of cells. In [142], a test statistic for the null hypothesis of a reference specimen is described as

$$\tilde{D} = E\left\{\left[F_S^{\text{ref}}(u) - \hat{F}_S^{\text{ref}}(u)\right]^2\right\}. \tag{4.57}$$

Here, $F_S^{\text{ref}}(u)$ denotes the *Empirical Cumulative Distribution Function* (ECDF) of the reconstructed complex amplitude of the biologoc cells. For retrieval of the sampling distribution of the test statistic, the $\hat{F}_S^{ref}(u)$ is calculated several times each with m different arbitrary test samples. Afterwards, the distribution of $\tilde{D}$ is computed and utilised as the *Criterion Discriminant Function* (CDF) proper for the null hypothesis. The test statistic for the input data is also given by

$$D = E\left\{\left[F_S^{ref}(u) - F_S^{inp}(u)\right]^2\right\} . \tag{4.58}$$

In order to conduct a hypothesis test, a null hypothesis H_0 and alternative hypothesis can be set up as follows:

$$H_0 : F_S^{ref}(u) = F_S^{inp}(u), \quad \text{for all } u \, H_0 : F_S^{ref}(u) \neq F_S^{inp}(u), \quad \text{for at least one } u .$$

In a last experiment, the statistical value is calculated by Monte Carlo techniques for the statistical decision to classify a specific specimen. Figure 4.68 depicts as a good working example the separability of sunflower and corn cells in a solution by combination of statistical sampling and Gabor transformed DHM hologram techniques.

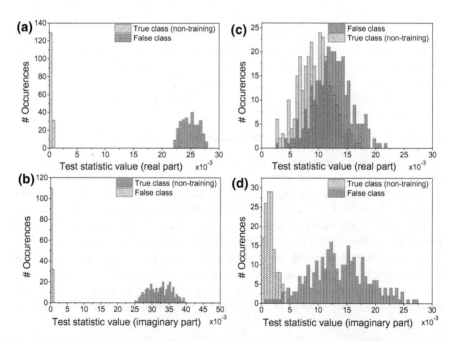

Fig. 4.68 Cell identification results utilizing statistical sampling theory and DHM on sunflower and corn cells in a solution: **a** Real and **b** imaginary parts of statistical sampling distributions of test statistic D for non-training true class input data against false class. **c** and **d** are the same as **a** and **b** but sampled from Gabor transformed digital holograms. The Gabor transform results in a separation of true and false classes

Combination of DHM and HOT for All-Optical Manipulation and Quantitative Phase Contrast of Cells

The wavelength of the trapping arm in the setup depicted in Fig. 4.32 can be chosen in near-IR region to minimize photo-induced damage due to high intensity of the tightly focused trapping beam. To achieve the best imaging resolution, one also has to choose the smallest wavelength possible for the imaging arm. This choice is restricted by the quantum efficiency of the CCD detector, the photo-damage on biological material at shorter wavelengths and the absorption by the microscope objective and other optical elements. It has to be considered that the imaging beam needs not to be illuminating the specimen at all times and can be controlled by an LCD shutter. In fact, in a single exposure on-axis setup only a short exposure time is needed to record a hologram. In the following, potential improvements in three areas are presented in more detail.

Automated Trapping of Cells Using Tracking Feedback

Even though 3D manipulation is possible with HOT in a fairly large axial range, it is practically difficult to control the trapped object if a 3D object detection mechanism is missing. In particular, creating twin traps depicted schematically in Fig. 2.13 requires precise measurement of the 3D position of cells in order to control the two independent traps accurately. These traps consist of a conventional trapping beam with an additional modulated convergent outer beam ring of different phase which passes the object outside and generates a trap from the backside after reflection on a mirror. DHM enables one to detect, track and determine the axial position of the cells with high-precision. Therefore, combining HOT and DHM systems allows for automation of the trapping process. For applications with micro-fluidics, an extended working distance can be achieved by using a low NA objective. In this case, holographic twin traps should be used for stable axial trapping. These traps can be controlled by the feedback from DHM-based cell tracking for precise axial positioning.

DHM can even be utilised for 3D tracking of several objects. With known properties of the microscope imaging system and digital holographic autofocusing, the relation between the propagation distance and the position of the specimen in object space enables the automated quantification of axial sample displacements. The calibration of the imaging system is performed by recording a set of stepwise defocused digital holograms, e.g., fixed micro beads on a glass carrier (see also Sect. 4.2.3 and Fig. 4.5). The determination of the focal image plane is performed by an autofocus algorithm that quantifies the image sharpness $fd(z)$ in dependence of the propagation distance z by summing up the logarithmically weighted and bandpass-filtered power spectra fd_{PS} of a reconstructed amplitude distribution $|O(x, y, z)|$ (see Sect. 4.2.1). In focus, pure phase objects like cells with negligible absorption occur with the least contrast in the amplitude distributions (autofocus criterion for phase objects) [143]. Thus, $fd_{PS} \rightarrow$ min is minimal whereas the corresponding quantitative phase contrast images are observed with optimised lateral and phase resolution. The automated obtained propagation distance z_{AF} of a sharp image of the sample is used to determine the relative axial object displacements Δg. The lateral cell position is determined from the automatically refocused phase contrast images. Therefore, from the unwrapped phase distributions within a region of interest (ROI), where the sample is located, the pixel coordinates of the phase maximum are determined after low-pass filtering (for illustration see dotted white boxes in Fig. 4.69a, b.

The time-dependent tracking of dynamic cellular displacements from reconstructed phase data of time-lapse sequences is performed by successive re-centering of the ROI to the coordinates of the preceding maximal phase value. The image scale is calibrated (e.g. by a USAF 1951 calibration test chart) to convert the resulting lateral displacement trajectories of the sample in direction from pixel coordinates to metric units. The combination of both, autofocus propagation distance and x-y-displacement trajectories, yields quantitative 3D object tracking.

The applicability of automated digital holographic 3D cell tracking is investigated by observation of the sedimentation process of three human red blood cells (RBCs, spherical spiky shaped echinocytes) in physiological solution with an experimental Mach–Zehnder interferometer setup as shown in Fig. 4.7b by application of frequency doubled Nd:YAG laser ($\lambda = 532$ nm) and a $63\times$ microscope lens ($NA = 0.75$). A series of 438 digital holograms ($\Delta t = 1$ s) has been recorded at fixed optical focus and afterwards is evaluated for 3D tracking. During the time dependent observation the cells perform fast dynamic motions. Thus, a permanent and robust numerical realignment of the focal position is required. Figure 4.69 shows the results of the experiment. Figure 4.69a exemplarily depicts a digital hologram of the three echinocytes 150 s after the start of digital holographic observation. The three tracked cells are denoted with A, B and C. Figure 4.69b presents sharply focused amplitude distributions for each cell obtained by application of digital holographic autofocus from the hologram in Fig. 4.69a. In Fig. 4.69c the corresponding quantitative phase distributions are depicted. The dotted boxes in Fig. 4.69b, c represent the ROIs that are applied for both, digital holographic autofocusing and dynamic xy-tracking. Figure 4.69d shows the resulting xy-trajectories of the RBCs obtained by the determination the coordinates with maximum phase contrast. In Fig. 4.69e the time dependence of the axial cell positions Δg that are obtained by digital holographic autofocusing are plotted. Figure 4.69f shows the 3D trajectories of the cells that result from combination of the data in Fig. 4.69d, e. Figure 4.69d–f demonstrate the reliability of digital holographic 3D tracking. Nearly 100 % of phase contrast images have been autofocused and reconstructed without unwrapping errors. Furthermore, the sedimentation process of the RBC due to gravitation is illustrated and quantified.

Probing Mechanical Properties of Cells

3D cell tracking using DHM allows a more extensive study of mechanical interaction of optical traps with cells by measuring optically induced cell deformations. The information can also be used for precise correction of the influence of internal flow. Due to the high flexibility of the HOT setup, 3D rotations (in-plane and out-of-plane) are possible for non-spherical objects ([144], not shown explicitly here). With special configurations even non-transparent objects can be rotated to some extend [145]. In combination with DHM, it is possible to perform high precision shape or volume measurements of the objects under investigation as they interact with optical traps to measure for instance the elasticity of cells. It is also possible to rotate the objects optically in micro-tomographic imaging applications to record different

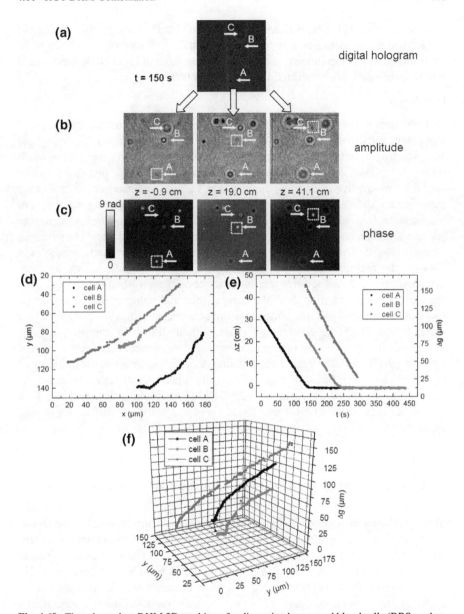

Fig. 4.69 Time dependent DHM 3D tracking of sedimenting human red blood cells (RBSs, spherical spiky shaped echinocytes). The tracked cells are denoted with A, B and C. **a**: exemplarily digital hologram of three echinocytes in different suspension layers 150 s after the begin of the experiment; **b**: sharply focused amplitude distributions for each cell obtained by application of digital holographic autofocus from the hologram in **a**; **c**: quantitative phase distributions obtained of the digital hologram in **a**; the dotted boxes in **b** and **c** represent the ROIs that are applied for digital holographic autofocus and dynamic xy-tracking; **d**: xy-trajectory of the RBCs obtained by determination the coordinates with maximum phase contrast; **e**: time dependence of the axial positions Δg of the cells obtained by digital holographic autofocusing; **f**: 3D trajectories obtained by combination of the data in **d** and **e**

exposures [146–148]. This is especially interesting for the tomographic imaging of mechanical micro-elements. For cells, unfortunately, this method has shown to be less suitable as the trap influences the inner arrangement of organelles, which could lead to inconsistencies between different measurements.

Conclusion

Cell classification and sorting is of interest in a variety of areas including isolation of pluripotent and differentiated stem cells for regenerative medicine, high throughput assays in drug development and separation of healthy embryos during in-vitro fertil-isation. It has been demonstrated that optical traps and 2D image processing can be used for non-invasive cell sorting applications. Cell discrimination by escape-force measurement shows a high sensitivity but only measures one integral parameter. For precise and robust cell identification, it is beneficial to combine this method with other cell identifications techniques like digital holography based statistical classification methods (also demonstrated in [124]). Extraction of cell location and boundaries in 3D space by real-time tracking algorithms using DHM therefore allows for auto-matic trapping and escape-force measurement of cells even when they move out of the focal plane. In addition, both the escape-force and holographic reconstruction data can be combined in a single framework to construct statistically optimal tests for identification. Therefore, combining the two methods allows for improved sensitiv-ity and specificity as a result of a more complete measurement set of cell attributes. Observations of the trap-organelle interaction or deformation under the influence of the optical force can be used as an indicator for a certain cell state or disease. As such interactions can are measurable with DHM in three-dimensions and with high precision, combination of HOT and DHM could be fruitful for further diagnostics applications.

4.12   Direct Laser Writing with Modified HOT-Setup

In the following the integrated multifunctional system setup utilised for the DHM-HOT-combination (Fig. 4.31) is characterised for the applicability of direct laser writ-ing (DLW) with multiple independent and dynamic focal spots. Therefore, instead of a micro-bead solution a negative photoresist (SU8 from MicroChem Corp., IP-L from Nanoscribe GmbH) is placed in a glass chamber and the infrared laser is exchanged by a visible cw-laser of 532 nm wavelength and ≈200 mW optical output power. Usually the efficiency for performing DLW is higher with a femtosecond laser, but the available SLMs (HoloEye Pluto) do not offer sufficient phase stability and therefore broadens the light pulse too much. First attempts of direct laser writing with cw-lasers and a conventional microscope with one fixed focus and a movable object holder have been reported by [149]. For multi spot 2PP direct laser writing only static spot distributions created by a fixed hologram plate have been utilised till now [150].

For the new experiments the optical system has to be re-optimised experimentally and in simulations for the new wavelength of 532 nm by exchanging the lenses in the HOT-beam path to those with visible anti-reflection coating. For the following experiments an oil-immersion microscope lens is applied (100× objective, Leica, $NA = 1.4$).

The search for the interface of glass substrate and photoresist is performed by recording the autocorrelation function of a low coherent light source (SLD or LED) and subsequent analysis of the correlogram with automated curve fitting. Thus, in contrast to a commercial microscope system (Zeiss), here the optical path of the reference beam is varied for the recording process (see correlogram of an LED in Fig. 4.70) offering more flexibility and detection of even weak interfaces with the drawback of a longer search duration.

Development of Written Structures

Usually, a special developer solution belongs to a photoresist. In the case of a negative resist, the developer dissolves the unexposed regions and in the case of a positive resist the developer dissolves the exposed regions of the resist. The photoresist IP-Dip is a negative resist and was designed especially for DiLL. The corresponding used developer is PGMEA. After the writing process the sample is put into the developer solution for at least 20 min. The sample is then washed with isopropanol which does not chemically affect the written structures. A fabrication of metal coated structures like nanoantennas requires further steps, e.g. evaporation, electrochemical deposition of metallic materials or a complete inversion in metal by a 4 or 5 step procedure [151].

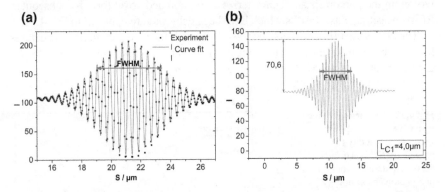

Fig. 4.70 a Recording of a LED's autocorrelation function for detecting the interface of glass substrate and photoresist by temporal scan in the reference path and curve fit of optical path difference. **b** Interpolated correlogram recorded with one camera image and off axis configuration of beams ("spatial scan")

(a)   **(b)**

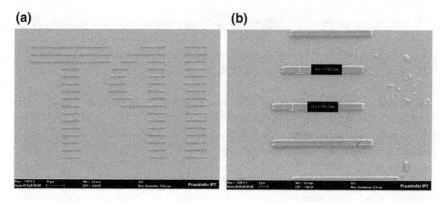

Fig. 4.71 SEM images of written lines by DLW with the DHM-HOT setup and a cw-laser of 532 nm in SU8 photoresist

Results of DLW with the HOT-System

Since the microscope object axes in the available setup is relatively imprecise with a repeatability of $\approx 2\,\mu m$ in the (x, y)-plane and $\approx 1\,\mu m$ in the z-axe, various attempts had to be performed in order to write lines in a multi-focal manner. The investigations have shown, that each generated spot should have an optical power of $>45\,mW$ to ensure that the photoresist is developed. Therefore, only two dynamic spots can be created with the available laser as the overall optical power coupled out of the MO is typically in the range of $[90–105]\,mW$. First, simple line pattern have been written in the photoresists to characterize resolution and voxel form by subsequent REM-measurements. Results which have been achieved are depicted in Fig. 4.71.

The photoresist IP-L was not successfully tested since the polymerised structures have not been sufficiently stable. Thus the TPA did not work here. The results of a characterisation of the SU8 photoresist are depicted in Fig. 4.72.

Fig. 4.72 The measured line widths versus laser power versus scan velocity for photoresist SU-8. The green curves are derived from the one-photon absorption model. The blue dotted lines are polynom fits from the corresponding two-photon absorption model [152]

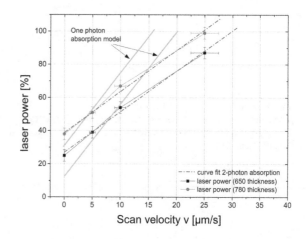

Conclusion and Outlook

The REM images show that the resolution of state of the art systems is not reached since the depicted lines have widths of currently ≈780 nm, thus roughly three times the width achievable with the Photonics Professional setup presented in Sect. 3.5 (typically 250 nm). In Fig. 4.71 it can be seen that the lines have a shape that is not very smooth, which may indicate that the intensity distribution in the focal points is not smooth either or does not have the shape of an ideal ellipsoid. It can be assumed that this is caused by the resting aberrations of the setup since a slight re-adjustment of the optical path changes the shape as well. Nevertheless, the aberrations could not totally be eliminated due to the unknown exact properties of the microscope objective in the simulation and the relatively simple catalogue lenses used for the setup. Iterative approaches by altering the phase distribution on the SLM in dependence of the line width have been conducted partly, but are highly time consuming.

Although a comprehensive, quantitative understanding of these results and the underlying complex chemical reactions are beyond the scope of the shown findings, it is assumed and reasonable that two-photon absorption is here equally responsible for the polymerisation since the photoresist SU8 exhibits lower but sufficient 2PP efficiency at the wavelength of 532 nm [153, 154]. Additionally, in [149] approximately analogue findings and even better line resolutions with 532 nm have been reported than with a *fs* pulsed laser at 780 nm. Subsequent to the introduction, this investigation opens up the potential for true 3D laser lithography with multi-focal and thus enhanced writing performance.

4.13 Nanoantenna Assisted Trapping

In current research, emphasis is often put on the micro-manipulation of small particles and tiny specimens. A auspicious approach is given by the micro manipulation on plasmonic nano-structured surfaces, that allows the disappearance of the diffraction limit for a specific achievable intensity gradient. Utilizing a surface with an array of preferably easy to fabricate simple optical nanoantennas, e.g. with each featuring the size of approximately $\lambda/2$, high local electromagnetic field amplifications can be achieved for allowing a flexible design of the potential energy distribution on the surface. This method is superior in trapping and moving nano particles (up to 20-times more efficiency compared to previous methods) as it has been demonstrated theoretically and experimentally in literature. The requirements on such a system are then significantly lower here in comparison to the previously suggested approaches [155]. An applied illumination with a low numerical aperture (NA) and a lower laser power has been reported to be sufficient for generating stable optical traps (usually 1 mW laser pointer) [155] that can additionally be supported by induced micro-fluidic flows [156].

Therefore in this section investigations on the application of gold (Au) nanoantenna arrays for increased efficient, 3D multi-purpose particle manipulation with decreased light power is explored. The utilisation of low laser light power implies that such a nanoantenna trapping system will be explicitly attractive for lab-on-a-chip technologies or biological applications aiming at reduced specimen photo damage.

Nanosphere Lithography

Since triangle structures are investigated for enhancing trapping capabilities, the general procedure for manufacturing these nanoantenna arrays is explained first.

In 1981 Fischer and Zingsheim were the first to publish an article about the preparation of nano-structures with the use of nano-sphere lithography (NSL) [157]. Since that time, many new approaches and improvements of established methods have been made. A respective overview is provided in [158]. Large-scale manufacturing of nano- or micrometre sized structures with methods like electron beam lithography, focused ion beam milling or direct laser writing yields high costs and poor throughput. Nano-sphere lithography, however, is a low cost technique, that enables easy fabrication of e.g. plasmonic structures of nano or micrometre size covering the hole substrate [159]. The general fabrication scheme is shown in Figs. 4.73 and 4.74. First, the solution containing the polystyrene beads is deposited by drop coating. A broad range of possible substrate materials like CaF_2, ZnS or Si can be used. In the process of drying, a monolayer of the spheres arrange themselves in a hexagonal order as shown in step 1. The polystyrene sphere monolayer now represents an almost perfect shadow mask for a metal evaporation. After erasing the spheres with e.g. adhesion tape, triangles arranged in a hexagonal array remain on the substrate. Although its major drawback lies the limited influence on the general structure shape, this fabrication method found its way to several different fields of applications like LED surface texturing [160] or substrate enhanced Raman scattering [161]. Beside that, nano-sphere lithography can be used for creating antennas suited for surface enhanced infrared spectroscopy covering the hole mid infrared range [162]. Simulations results according to the hexagonally arranged, triangular structures have been performed with an FDTD-software (Lumerical).

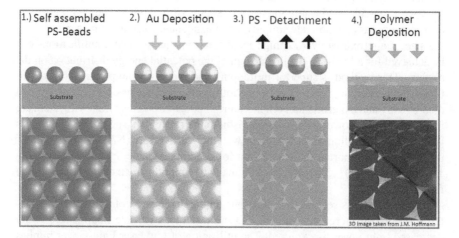

Fig. 4.73 Scheme for illustration of the nano-sphere lithography fabrication process

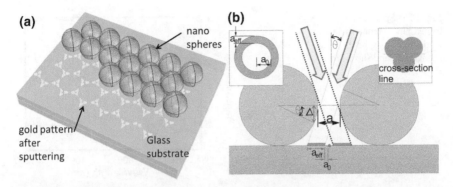

Fig. 4.74 **a** 3D-illustration of adhered nano-spheres on the substrate which represent different masks according to the projection by the angle of gold deposition. **b** Geometric illustration showing dependency of angle of gold deposition and antenna dimensions

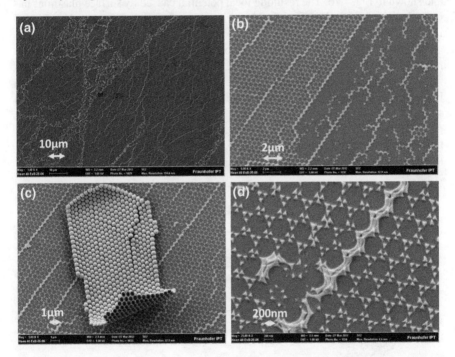

Fig. 4.75 SEM-image of 50 nm sized gold "trimers" measurements of different magnification (**a** 1000x, **b** 5000x, **c** 5000x, **d** 25000x). In **c** a resting part of the nano-sphere monolayer can be seen

The fabricated 50 nm sized gold "trimers" are depicted in Fig. 4.75. Here, most of the surface is homogeneous, but also non perfect surface parts are shown with high magnification.

Fig. 4.76 Spectrum of the
trimer nanoantenna surface
made by nano-sphere
lithography and shown in
Fig. 4.75

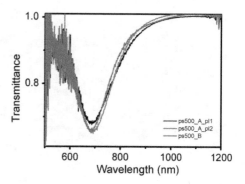

The spectrum of the antenna array is measured with an FTIR-system (Bruker)
and shown in Fig. 4.76. The resulting local potential caused by surface plasmons has
been simulated with a FDTD-software environment (Lumerical) and is plotted in 3D
in Fig. 4.77.

In investigations on the trap force as explained in Sect. 4.11 based on the maximum speed of a trapped particle in a fluid, the trapping force with the nanoantenna
surface turned out to be 12% ($\pm$3%) higher with 4 μm beads and with $\lambda = 1060$ nm
in comparison to normal glass slides used in microscopy as proven by empirical
studies shown in Fig. 4.78.

The results of the statistical analysis are summarised by the key parameters of
the Gaussian fit: Mean value (centre): (0.5009 ± 0.0016)pN; Standard deviation:
(0.0366 ± 0.0022)pN; FWHM: (0.0862 ± 0.0054)pN.

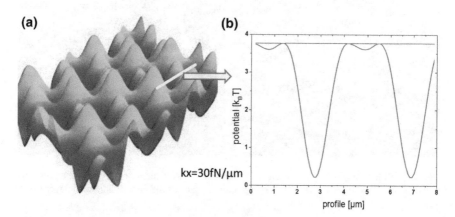

Fig. 4.77 a 3D view of an optical binding energy trap computed for a 300 nm polystyrene bead on
a surface of gold triangle nanoantenna arrays. **b** Trapping potential along an indicated profile of a
4 μm polystyrene sphere bead under *p* polarisation

Fig. 4.78 Histogram of determined trap forces including Gaussian parameter fit. Here the trap force increased $(12 \pm 3)\%$ with $4\,\mu$m micro beads in comparison to a standard glass surface

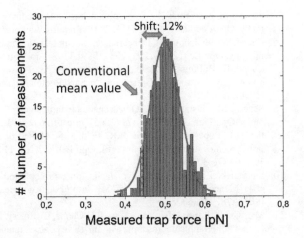

Conclusion and Outlook

It is demonstrated that the enhancement of the trapping capability is slightly increased with the application of nanoantenna surface which causes higher electromagnetic fields locally. For further experiments and more in-depth investigations under the availability of more precise xy-stage, empirically obtained plasmonic optical trapping "phase diagrams" are envisaged to characterise the trapping response of the nanoantenna arrays thoroughly as a function of input power, wavelength, particle diameter, polarisation and NA similar to [155]. Using these diagrams, parameters can then be chosen, employing the degrees-of-freedom of the input light, to engineer specific trapping tasks including complex single particle trapping, and manipulation, trapping and manipulation of two- and three dimensional particle clusters and particle sorting. Better enhancement factors have been reported for Bowtie antenna arrays when using low aperture traps [163]. But the advantage of the trimers here is an easy fabrication process. Other lithographic methods have not been available in the framework of this work.

References

1. Poon, T.-C., Motamedi, M.: Optical/digital incoherent image processing for extended depth of field. Appl. Opt. **26**(21), 4612–4615 (1987). https://doi.org/10.1364/AO.26.004612
2. Dowski, E.R., Jr., Cathey, W.T.: Extended depth of field through wave-front coding. Appl. Opt. **34**(11), 1859–1866 (1995). https://doi.org/10.1364/AO.34.001859
3. Schnars, U., Jueptner, W.: Direct recording of holograms by a CCD target and numerical reconstruction. Appl. Opt. **33**(2), 179–181 (1994). https://doi.org/10.1364/AO.33.000179
4. Yamaguchi, I., Zhang, T.: Phase-shifting digital holography. Opt. Lett. **22**(16), 1268–1270 (1997). https://doi.org/10.1364/OL.22.001268

5. Zhang, T., Yamaguchi, I.: Three-dimensional microscopy with phase-shifting digital holography. Opt. Lett. **23**(15), 1221–1223 (1998). https://doi.org/10.1364/OL.23.001221
6. Cuche, E., Bevilacqua, F., Depeursinge, C.: Digital holography for quantitative phase-contrast imaging. Opt. Lett. **24**(5), 291–293 (1999). https://doi.org/10.1364/OL.24.000291
7. Cuche, E., Ruffieux, P., Poscio, P., Depeursinge, C.: Phase contrast microscopy with digital holography. Advances in Optical Imaging and Photon Migration. Optical Society of America (1998), ATuA3
8. Sebesta, M., Gustafsson, M.: Object characterization with refractometric digital Fourier holography. Opt. Lett. *30*(5), 471–473 (2005). https://doi.org/10.1364/OL.30.000471
9. Ikeda, T., Popescu, G., Dasari, R.R., Feld, M.S.: Hilbert phase microscopy for investigating fast dynamics in transparent systems. Opt. Lett. **30**(10), 1165–1167 (2005). https://doi.org/10.1364/OL.30.001165
10. Smigielski, P., Faupel, M., Grzymala, R.: Imagerie et photonique: Pour les Sciences du Vivant et la Médecine. FontisMedia (2004). http://books.google.de/books?id=iH1kPQAACAAJ. ISBN 9782884760058
11. Marquet, P., Rappaz, B., Magistretti, P.J., Cuche, E., Emery, Y., Colomb, T., Depeursinge, C.: Digital holographic microscopy: a noninvasive contrast imaging technique allowing quantitative visualization of living cells with subwavelength axial accuracy. Opt. Lett. **30**(5), 468–470 (2005). https://doi.org/10.1364/OL.30.000468
12. Dubois, F., Debeir, O., Kiss, R., Decaestecker, C., Van Ham, P., Yourassowsky, C., Monnom, O., Legros, J.-C.: Digital holographic microscopy for the three-dimensional dynamic analysis of in vitro cancer cell migration. J. Biomed. Opt. **11**(5), 054032–054032-5 (2006). https://doi.org/10.1117/1.2357174, ISBN 1083–3668
13. Lue, N., Popescu, G., Ikeda, T., Dasari, R.R., Badizadegan, K., Feld, M.S.: Live cell refractometry using microfluidic devices. Opt. Lett. **31**(18), 2759–2761 (2006). https://doi.org/10.1364/OL.31.002759
14. Marquet, P., Rappaz, B., Charrièrec, F., Emery, Y., Depeursinge, C., Magistretti, P.: Analysis of cellular structure and dynamics with digital holographic microscopy. Biophotonics 2007: Optics in Life Science, vol. 6633. Optical Society of America (2007)
15. Ferraro, P., Nicola, S.D., Coppola, G., Finizio, A., Alfieri, D., Pierattini, G.: Controlling image size as a function of distance and wavelength in Fresnel-transform reconstruction of digital holograms. Opt. Lett. **29**(8), 854–856 (2004). https://doi.org/10.1364/OL.29.000854
16. Ferraro, P., Nicola, S.D., Finizio, A., Coppola, G., Grilli, S., Magro, C., Pierattini, G.: Compensation of the inherent wave front curvature in digital holographic coherent microscopy for quantitative phase-contrast imaging. Appl. Opt. **42**(11), 1938–1946 (2003). https://doi.org/10.1364/AO.42.001938
17. Colomb, T., Cuche, E., Charrière, F., Kühn, J., Aspert, N., Montfort, F., Marquet, P., Depeursinge, C.: Automatic procedure for aberration compensation in digital holographic microscopy and applications to specimen shape compensation. Appl. Opt. **45**(5), 851–863 (2006). https://doi.org/10.1364/AO.45.000851
18. Javidi, B., Tajahuerce, E.: Three-dimensional object recognition by use of digital holography. Opt. Lett. **25**(9), 610–612 (2000). https://doi.org/10.1364/OL.25.000610
19. Kim, D., Javidi, B.: Distortion-tolerant 3-D object recognition by using single exposure on-axis digital holography. Opt. Express **12**(22), 5539–5548 (2004). https://doi.org/10.1364/OPEX.12.005539
20. Dubois, F., Monnom, O., Yourassowsky, C., Legros, J.-C.: Border processing in digital holography by extension of the digital hologram and reduction of the higher spatial frequencies. Appl. Opt. **41**(14), 2621–2626 (2002). https://doi.org/10.1364/AO.41.002621
21. Klysubun, P., Indebetouw, G.: A posteriori processing of spatiotemporal digital microholograms. J. Opt. Soc. Am. A **18**(2), 326–331 (2001). https://doi.org/10.1364/JOSAA.18.000326

22. Yu, L., Cai, L.: Iterative algorithm with a constraint condition for numerical reconstruction of a three-dimensional object from its hologram. J. Opt. Soc. Am. A **18**(5), 1033–1045 (2001). https://doi.org/10.1364/JOSAA.18.001033

23. Li, W., Loomis, N.C., Hu, Q., Davis, C.S.: Focus detection from digital in-line holograms based on spectral l1 norms. J. Opt. Soc. Am. A **24**(10), 3054–3062 (2007). https://doi.org/10.1364/JOSAA.24.003054

24. McElhinney, CP., McDonald, JB., Castro, A., Frauel, Y., Javidi, B., Naughton, TJ.: Depth-independent segmentation of macroscopic three-dimensional objects encoded in single perspectives of digital holograms. Opt. Lett. **32**(10), 1229–1231 (2007). https://doi.org/10.1364/OL.32.001229

25. Paturzo, M., Ferraro, P.: Creating an extended focus image of a tilted object in Fourier digital holography. Opt. Express **17**(22), 20546–20552 (2009). https://doi.org/10.1364/OE.17.020546

26. Kim, T., Poon, T.-C., Indebetouw, G.: Depth detection and image recovery in remote sensing by optical scanning holography. Opt. Eng. **41**(6), 1331–1338 (2002). https://doi.org/10.1117/1.1476939, ISBN 0091–3286

27. Dubois, F., Schockaert, C., Callens, N., Yourassowsky, C.: Focus plane detection criteria in digital holography microscopy by amplitude analysis. Opt. Express **14**(13), 5895–5908 (2006). https://doi.org/10.1364/OE.14.005895

28. Sung, Y., Choi, W., Fang-Yen, C., Badizadegan, K., Dasari, R.R., Feld, M.S.: Optical diffraction tomography for high resolution live cell imaging. Opt. Express **17**(1), 266–277 (2009). https://doi.org/10.1364/OE.17.000266

29. Hong, S.-H., Jang, J.-S., Javidi, B.: Three-dimensional volumetric object reconstruction using computational integral imaging. Opt. Express **12**(3), 483–491 (2004). https://doi.org/10.1364/OPEX.12.000483

30. Mann, C., Yu, L., Lo, C.-M., Kim, M.: High-resolution quantitative phase-contrast microscopy by digital holography. Opt. Express **13**(22), 8693–8698 (2005). https://doi.org/10.1364/OPEX.13.008693

31. Martínez-León, L., Pedrini, G., Osten, W.: Applications of short-coherence digital holography in microscopy. Appl. Opt. **44**(19), 3977–3984 (2005). https://doi.org/10.1364/AO.44.003977

32. Charrière, F., Kühn, J., Colomb, T., Montfort, F., Cuche, E., Emery, Y., Weible, K., Marquet, P., Depeursinge, C.: Characterization of microlenses by digital holographic microscopy. Appl. Opt. **45**(5), 829–835 (2006). https://doi.org/10.1364/AO.45.000829

33. Sun, Y., Duthaler, S., Nelson, B.J.: Autofocusing in computer microscopy: selecting the optimal focus algorithm. Microsc. Res. Tech. **65**(3), 139–149 (2004). https://doi.org/10.1002/jemt.20118, ISSN 1097–0029

34. Wolf, F., Geley, S.: A simple and stable autofocusing protocol for long multidimensional live cell microscopy. J. Microsc. **221**(1), 72–77 (2006). https://doi.org/10.1111/j.1365-2818.2006.01538.x, ISSN 1365–2818

35. Liebling, M., Unser, M.: Autofocus for digital Fresnel holograms by use of a Fresnelet-sparsity criterion. J. Opt. Soc. Am. A **21**(12), 2424–2430 (2004)

36. Remmersmann, C., Stürwald, S., Kemper, B., Langehanenberg, P., von Bally, G.: Phase noise optimization in temporal phase-shifting digital holography with partial coherence light sources and its application in quantitative cell imaging. Appl. Opt. **48**(8), 1463–1472 (2009). https://doi.org/10.1364/AO.48.001463

37. Stürwald, S., Kemper, B., Remmersmann, C., Langehanenberg, P., von Bally, G : Application of light emitting diodes in digital holographic microscopy. Version 2008. https://doi.org/10.1117/12.781186

38. Forbes, G.W.: Shape specification for axially symmetric optical surfaces. Opt. Express **15**(8), 5218–5226 (2007). https://doi.org/10.1364/OE.15.005218

39. Kehtarnavaz, N., Oh, H.-J.: Development and real-time implementation of a rule-based auto-focus algorithm. Real-Time Imaging **9**(3), 197–203 (2003). https://doi.org/10.1016/S1077-2014(03)00037-8, ISSN 10772014

40. Geusebroek, J.-M., Cornelissen, F., Smeulders, A.W., Geerts, H.: Robust autofocus-ing in microscopy. Cytometry **39**(1), 1–9 (2000). http://dx.doi.org/10.1002/(SICI)1097-0320(20000101)39:11::AID-CYTO23.0.CO;2-J, ISSN 1097–0320

41. Gabarda, S., Cristóbal, G.: No-reference image quality assessment through the von Mises distribution. J. Opt. Soc. Am. A **29**(10), 2058–2066 (2012). https://doi.org/10.1364/JOSAA.29.002058

42. Yibin, T., Zhihai, X., Huajun, F., Ruichun, D.: Autofocus based on skin detection **32**(1), 66–60 (2013)

43. Chern, N.N.K., Neow, P.A., Ang, V.M.H.: Practical issues in pixel-based autofocusing for machine vision. In: IEEE International Conference on Robotics and Automation. Proceedings 2001 ICRA, vol. 3, 2791–2796 (2001). ISSN 1050–4729

44. Firestone, L., Cook, K., Culp, K., Talsania, N., Preston, K.: Comparison of autofocus methods for automated microscopy. Cytometry **12**(3), 195–206 (1991). https://doi.org/10.1002/cyto.990120302, ISSN 1097–0320

45. Bravo-Zanoguera, M., Massenbach, Bv., Kellner, A.L., Price, J.H.: High-performance auto-focus circuit for biological microscopy. Rev. Sci. Instrum. **69**(11), 3966–3977 (1998). https://doi.org/10.1063/1.1149207

46. Agero, U., Mesquita, L.G., Neves, B.R.A., Gazzinelli, R.T., Mesquita, O.N.: Defocusing microscopy. Microsc. Res. Tech. **65**(3), 159–165 (2004). https://doi.org/10.1002/jemt.20117, ISSN 1097–0029

47. Langehanenberg, P., Kemper, B., Dirksen, D., von Bally, G.: Autofocusing in digital holo-graphic phase contrast microscopy on pure phase objects for live cell imaging. Appl. Opt. **47**(19), D176–D182 (2008). https://doi.org/10.1364/AO.47.00D176

48. Langehanenberg, P., Kemper, B., von Bally, G.: Autofocus algorithms for digital holographic microscopy. Biophotonics 2007: Optics in Life Science, pp. 613–633. Optical Society of America (2007)

49. Schofield, M.A., Zhu, Y.: Fast phase unwrapping algorithm for interferometric applications. Opt. Lett. **28**(14), 1194–1196 (2003). https://doi.org/10.1364/OL.28.001194

50. Valdecasas, A.G., Marshall, D., Becerra, J.M., Terrero, J.J.: On the extended depth of focus algorithms for bright field microscopy. Micron **32**(6), 559–569 (2001). https://doi.org/10.1016/S0968-4328(00)00061-5, ISSN 0968–4328

51. Press, W.H., Teukolsky, S.A., Vetterling, W.T., Flannery, B.P.: Numerical Recipes in C. The Art of Scientific Computing, 2nd edn. Cambridge University Press, New York (1992). ISBN 0–521–43108–5

52. He, J., Zhou, R., Hong, Z.: Modified fast climbing search auto-focus algorithm with adaptive step size searching technique for digital camera. IEEE Trans. Consum. Electron. **49**(2), 257–262 (2003)

53. Faure, H., Lemieux, C.: Generalized Halton sequences in 2008: a comparative study. ACM Trans. Model. Comput. Simul. **19**(4), 15:1–15:31 (2009). https://doi.org/10.1145/1596519.1596520, ISSN 1049–3301

54. Halton, J.H.: Algorithm 247: radical-inverse quasi-random point sequence. Commun. ACM **7**(12), 701–702 (1964). https://doi.org/10.1145/355588.365104, ISSN 00010782

55. Niederreiter, H.: Random Number Generation and Quasi-Monte Carlo Methods: Regional Research Conference Selected Papers. Society for Industrial and Applied Mathematics, Philadelphia (1992). ISBN 9780898712957

56. Kuipers, L., Niederreiter, H.: Uniform Distribution of Sequences. Dover Books on Mathematics. Dover Publications, Mineola (2006). ISBN 9780486450193

57. Colomb, T., Montfort, F., Kühn, J., Aspert, N., Cuche, E., Marian, A., Charrière, F., Bourquin, S., Marquet, P., Depeursinge, C.: Numerical parametric lens for shifting, magnification, and complete aberration compensation in digital holographic microscopy. J. Opt. Soc. Am. A **23**(12), 3177–3190 (2006). https://doi.org/10.1364/JOSAA.23.003177

58. Dubois, F., Joannes, L., Legros, J.-C.: Improved three-dimensional imaging with a digital holography microscope with a source of partial spatial coherence. Appl. Opt. **38**(34), 7085–7094 (1999). https://doi.org/10.1364/AO.38.007085

59. Dubois, F., Requena, M.-L.N., Minetti, C., Monnom, O., Istasse, E.: Partial spatial coherence effects in digital holographic microscopy with a laser source. Appl. Opt. **43**(5), 1131–1139 (2004). https://doi.org/10.1364/AO.43.001131

60. Mehta, D.S., Saxena, K., Dubey, S.K., Shakher, C.: Coherence characteristics of light-emitting diodes. J. Lumin. **130**(1), 96–102 (2010). https://doi.org/10.1016/j.jlumin.2009.07.027, ISSN 0022–2313

61. Kemper, B., Kosmeier, S., Langehanenberg, P., Przibilla, S., Remmersmann, C., Stürwald, S., von Bally, G.: Application of 3D tracking, LED illumination and multi-wavelength techniques for quantitative cell analysis in digital holographic microscopy. Version 2009. https://doi.org/10.1117/12.808392

62. Takeda, M., Ina, H., Kobayashi, S.: Fourier-transform method of fringe-pattern analysis for computer-based topography and interferometry. J. Opt. Soc. Am. **72**(1), 156–160 (1982). https://doi.org/10.1364/JOSA.72.000156

63. Kreis, T.: J. Opt. Soc. Am. A **6**, 847–855

64. Chavel, P., Lowenthal, S.: Noise and coherence in optical image processing. II. Noise fluctuations. J. Opt. Soc. Am. **68**(6), 721 (1978). https://doi.org/10.1364/JOSA.68.000721

65. Dubois, F., Yourassowski, C.: Method and device for obtaining a sample with three dimensional microscopy (2006)

66. Liebling, M., Unser, M.: Autofocus for digital Fresnel holograms by use of a Fresnelet-sparsity criterion. J. Opt. Soc. Am. A **21**(12), 2424–2430 (2004). https://doi.org/10.1364/JOSAA.21.002424

67. Kemper, B., Stuerwald, S., Remmersmann, C., Langehanenberg, P., von Bally, G.: Characterisation of light emitting diodes (LEDs) for application in digital holographic microscopy for inspection of micro and nanostructured surfaces. Opt. Lasers Eng. **46**(7), 499–507 (2008). https://doi.org/10.1016/j.optlaseng.2008.03.007, ISSN 0143–8166

68. Kreis, T.: Holographic Interferometry: Principles and Methods. Akademie Verlag Series in Optical Metrology. Akademie Verlag, Berlin (1996). http://books.google.de/books?id=qfJRAAAAMAAJ. ISBN 9783055016448

69. Javidi, B., Do, C.M., Hong, S.-H., Nomura, T.: Multi-spectral holographic three-dimensional image fusion using discrete wavelet transform. J. Disp. Technol. **2**(4), 411–417 (2006). https://doi.org/10.1109/JDT.2006.885156, ISSN 1551–319X

70. Kuhn, K.J.: Laser Engineering. Prentice Hall Books, Upper Saddle River (1998). http://books.google.de/books?id=C-WVQgAACAAJ. ISBN 9780023669217

71. Siegman, A.E.: Lasers. University Science Books (1986). http://books.google.de/books?id=1BZVwUZLTkAC. ISBN 9780935702118

72. Bouma, B.: Handbook of Optical Coherence Tomography. Taylor & Francis, Boca Raton (2001). http://books.google.de/books?id=3r3zgdeO8XgC. ISBN 9780824705589

73. Dai, G., Koenders, L., Pohlenz, F., Dziomba, T., Danzebrink, H.-U.: Accurate and traceable calibration of one-dimensional gratings. Meas. Sci. Technol. **16**(6), 1241–1249 (2005). https://doi.org/10.1088/0957-0233/16/6/001, ISSN 0957–0233

74. Weckenmann, A., Gawande, B.: Koordinatenmesstechnik: Flexible Messstrategien für Mass, Form und Lage. Carl Hanser (1999). http://books.google.de/books?id=Fnn2AQAACAAJ. ISBN 9783446179912

75. Wiedenhöfer, T., Weckenmann, A.: Modelling measurement processes in nanometrology - aspects on specific sources of uncertainty. In: Proceedings of the 6th EUSPEN International Conference, Vol. I, 402–405 (2006)

76. Marinello, F., Bariani, P., Carmignato, S., Savio, E.: Geometrical modelling of scanning probe microscopes and characterization of errors. Meas. Sci. Technol. **20**(8), 084013 (2009). http://stacks.iop.org/0957-0233/20/i=8/a=084013

77. Marinello, F., Savio, E., Bariani, P., Carmignato, S.: Coordinate metrology using scanning probe microscopes. Meas. Sci. Technol. **20**(8), 084002 (2009). http://stacks.iop.org/0957-0233/20/i=8/a=084002

78. Zhang, H., Huang, F., Higuchi, T.: Dual unit scanning tunneling microscope-atomic force microscope for length measurement based on reference scales. J. Vac. Sci. Technol. B **15**(4), 780–784 (1997). https://doi.org/10.1116/1.589408

79. Jo/rgensen, J.F., Madsen, L.L., Garnaes, J., Carneiro, K., Schaumburg, K.: Calibration, drift elimination, and molecular structure analysis. J. Vac. Sci. Technol. B **12**(3), 1698–1701 (1994). https://doi.org/10.1116/1.587266

80. http://www.microscopyu.com/articles/optics/cfintro.html

81. Seward, G.: Tutorial texts in optical engineering. Optical Design of Microscopes, vol. TT88. SPIE Press, Bellingham and Wash (2010). ISBN 0819480959

82. Dufresne, E.R., Spalding, G.C., Dearing, M.T., Sheets, S.A., Grier, D.G.: Computer generated holographic optical tweezer arrays. Rev. Sci. Instrum. **72**(3), 1810 (2001). https://doi.org/10.1063/1.1344176

83. Hayasaki, Y., Itoh, M., Yatagai, T., Nishida, N.: Nonmechanical optical manipulation of microparticle using spatial light modulator. Opt. Rev. **6**(1), 24–27 (1999). https://doi.org/10.1007/s10043-999-0024-5, ISSN 1340–6000

84. Hossack, W., Theofanidou, E., Crain, J., Heggarty, K., Birch, M.: High-speed holographic optical tweezers using a ferroelectric liquid crystal microdisplay. Opt. Express **11**(17), 2053 (2003). https://doi.org/10.1364/OE.11.002053

85. Martín-Badosa, E., Montes-Usategui, M., Carnicer, A., Andilla, J., Pleguezuelos, E., Juvells, I.: Design strategies for optimizing holographic optical tweezers set-ups. J. Opt. A: Pure Appl. Opt. **9**(8), S267–S277 (2007). https://doi.org/10.1088/1464-4258/9/8/S22, ISSN 1464–4258

86. Gerchberg, R.W., Saxton, W.O.: A practical algorithm for the determination of phase from image and diffraction plane pictures. Optik **35**, 237 (1971)

87. Di Leonardo, R., Ianni, F., Ruocco, G.: Computer generation of optimal holograms for optical trap arrays. Opt. Express **15**(4), 1913–1922 (2007). https://doi.org/10.1364/OE.15.001913

88. Fukuchi, N., Biqing, YE, Igasaki, Y., Yoshida, N., Kobayashi, Y., Hara, T.: Oblique-incidence characteristics of a parallel-aligned nematic-liquid-crystal spatial light modulator. Opt. Rev. **12**(5), 372 (2005). https://doi.org/10.1007/s10043-005-0372-8

89. Mitutoyo Deutschland GmbH: Optik-Glossar. 09.04.2008. http://www2.mitutoyo.de/fileadmin/user_upload/pdf/prospekte/optische_messgeraete/Optik-Glossar_kl.pdf

90. Zwick, S.: Flexible Mikromanipulation durch räumliche Lichtmodulation in der Mikroskopie. Dissertation, Universität Stuttgart, Stuttgart, 19.05.2010

91. Demtröder, W.: Laser spectroscopy: basic concepts and instrumentation; with 91 problems and hints for solution, vol. 3. Springer, Berlin (2002). ISBN 3540652256

92. Thorlabs: Optical Tweezer Kit (2012). http://www.thorlabs.de/catalogpages/V21/1825.PDF

93. Martín-Badosa, E., Montes-Usategui, M., Carnicer, A., Andilla, J., Pleguezuelos, E., Juvells, I.: Design strategies for optimizing holographic optical tweezers set-ups. J. Opt. A: Pure Appl. Opt. **9**(8), S267–S277 (2007). ISSN 1464–4258

94. Bowman, R., D'Ambrosio, V., Rubino, E., Jedrkiewicz, O., Di Trapani, P., Padgett, M.J.: Optimisation of a low cost SLM for diffraction efficiency and ghost order suppression. Eur. Phys. J. Special Top. **199**(1), 149–158 (2011). https://doi.org/10.1140/epjst/e2011-01510-4, ISSN 1951–6355

95. Grier, D.G.: A revolution in optical manipulation. Nature **424**(6950), 810–816 (2003). https://doi.org/10.1038/nature01935

96. Ccaron, T., Mazilu, M., Dholakia, K.: In situ wavefront correction and its application to micromanipulation. Nat. Photonics **4**(6), 388–394 (2010). https://doi.org/10.1038/nphoton.2010.85

97. Dholakia, K., Cizmar, T.: Shaping the future of manipulation. Nat. Photonics **5**(6), 335–342 (2011). https://doi.org/10.1038/nphoton.2011.80

98. Dufresne, E.R., Grier, D.G.: Optical tweezer arrays and optical substrates created with diffractive optics. Rev. Sci. Instrum. **69**(5), 1974–1977 (1998). https://doi.org/10.1063/1.1148883

99. Curtis, J.E., Koss, B.A., Grier, D.G.: Dynamic holographic optical tweezers. Opt. Commun. **207**(1–6), 169–175 (2002). https://doi.org/10.1016/S0030-4018(02)01524-9, ISSN 0030–4018

100. Yeh, P., Gu, C.: Optics of Liquid Crystal Displays. Wiley Series in Pure and Applied Optics. Wiley, New York (2010). http://books.google.de/books?id=0XhtwBpMtA8C. ISBN 9780470181768

101. Reicherter, M., Wagemann, E.U., Haist, T., Tiziani, H.J.: Optical particle trapping with computer-generated holograms written on a liquid-crystal display. Opt. Lett. **24**(9), 608–610 (1999). https://doi.org/10.1364/OL.24.000608

102. van der Horst, A., Forde, N.R.: Calibration of dynamic holographic optical tweezers for force measurements on biomaterials. Opt. Express **16**(25), 20987–21003 (2008). https://doi.org/10.1364/OE.16.020987

103. Moffitt, J.R., Chemla, Y.R., Smith, S.B., Bustamante, C.: Recent advances in optical tweezers. Annu. Rev. Biochem. **77**(1), 205–228 (2008). https://doi.org/10.1146/annurev.biochem.77.043007.090225. PMID: 18307407

104. van der Horst, A., Downing, B.P.B., Forde, N.R.: Position and intensity modulations in holographic optical traps created by a liquid crystal spatial light modulator. Advances in Imaging. Optical Society of America (2009), OMB3

105. McLane, L.T., Carroll, K.M., Scrimgeour, J., Bedoya, M.D., Kramer, A., Curtis, J.E.: Force measurements with a translating holographic optical trap. Version 2010. https://doi.org/10.1117/12.863549

106. Padgett, M.: The effect of external forces on discrete motion within holographic optical tweezers. Opt. Express **15**(26), 18268–18274 (2007). http://eprints.gla.ac.uk/29118/

107. Persson, M., Engström, D., Frank, A., Backsten, J., Bengtsson, J., Goksör, M.: Minimizing intensity fluctuations in dynamic holographic optical tweezers by restricted phase change. Opt. Express **18**(11), 11250–11263 (2010). https://doi.org/10.1364/OE.18.011250

108. McKnight, D.J.: Continuous view of dc-balanced images on a ferroelectric liquid-crystal spatial light modulator. Opt. Lett. **19**(18), 1471–1473 (1994). https://doi.org/10.1364/OL.19.001471

109. Moore, J.R., Collings, N., Crossland, W.A., Davey, A.B., Evans, M., Jeziorska, A.M., Komarcevic, M., Parker, R.J., Wilkinson, T.D., Xu, H.: The silicon backplane design for an LCOS polarization-insensitive phase hologram SLM. IEEE Photonics Technol. Lett. **20**(1), 60–62 (2008). https://doi.org/10.1109/LPT.2007.912690, ISSN 1041–1135

110. Hermerschmidt, A., Osten, S., Krüger, S., Blümel, T.: Wave front generation using a phase-only modulating liquid-crystal-based micro-display with HDTV resolution. Version 2007. https://doi.org/10.1117/12.722891

111. Moreno, I., Lizana, A., Márquez, A., Iemmi, C., Fernández, E., Campos, J., Yzuel, M.J.: Time fluctuations of the phase modulation in a liquid crystal on silicon display: characterization and effects in diffractive optics. Opt. Express **16**(21), 16711–16722 (2008). https://doi.org/10.1364/OE.16.016711

112. Li, F., Mukohzaka, N., Yoshida, N., Igasaki, Y., Toyoda, H., Inoue, T., Kobayashi, Y., Hara, T.: Phase modulation characteristics analysis of optically-addressed parallel-aligned nematic liquid crystal phase-only spatial light modulator combined with a liquid crystal display. Opt. Rev. **5**(3), 174–178 (1998). https://doi.org/10.1007/s10043-998-0174-x, ISSN 1340–6000

113. Berg-Soerensen, K., Oddershede, L., Florin, E.-L., Flyvbjerg, H.: Unintended filtering in a typical photodiode detection system for optical tweezers. J. Appl. Phys. **93**(6), 3167–3176 (2003). https://doi.org/10.1063/1.1554755

114. Tolić-Nørrelykke, S.F., Schäffer, E., Howard, J., Pavone, F.S., Jülicher, F., Flyvbjerg, H.: Calibration of optical tweezers with positional detection in the back focal plane. Rev. Sci. Instrum. **77**(10), 103101 (2006). https://doi.org/10.1063/1.2356852

115. Widdershoven, I., Donker, R.L., Spaan, H.A.M.: Realization and calibration of the "Isara 400" ultra-precision CMM. J. Phys.: Conf. Ser. **311**(1), 012002 (2011). http://stacks.iop.org/1742-6596/311/i=1/a=012002

116. Manske, E., Hausotte, T., Mastylo, R., Machleidt, T., Franke, K.-H., Jäger, G.: New applications of the nanopositioning and nanomeasuring machine by using advanced tactile and non-tactile probes. Meas. Sci. Technol. **18**(2), 520 (2007). http://stacks.iop.org/0957-0233/18/i=2/a=S27

117. Weckenmann, A., Peggs, G., Hoffmann, J.: Probing systems for dimensional micro- and nano-metrology. Meas. Sci. Technol. **17**(3), 504 (2006). http://stacks.iop.org/0957-0233/17/i=3/a=S08

118. Wong, D.W.K., Chen, G.: Redistribution of the zero order by the use of a phase checkerboard pattern in computer generated holograms. Appl. Opt. **47**(4), 602–610 (2008). https://doi.org/10.1364/AO.47.000602

119. Dudley, A., Vasilyeu, R., Belyi, V., Khilo, N., Ropot, P., Forbes, A.: Controlling the evolution of nondiffracting speckle by complex amplitude modulation on a phase-only spatial light modulator. Opt. Commun. **285**(1), 5–12 (2012). https://doi.org/10.1016/j.optcom.2011.09.004, ISSN 0030–4018

120. Cojocaru, E.: Mathieu functions computational toolbox implemented in Matlab 2008. arXiv:0811.1970. – Forschungsbericht. – Comments: 20pages,0figures,6tables

121. MacDonald, M.P.: Creation and manipulation of three-dimensional optically trapped structures. Science **296**(5570), 1101–1103 (2002). https://doi.org/10.1126/science.1069571, ISSN 00368075

122. Dudley, A., Forbes, A.: From stationary annular rings to rotating Bessel beams. J. Opt. Soc. Am. A **29**(4), 567–573 (2012). https://doi.org/10.1364/JOSAA.29.000567

123. DaneshPanah, M., Javidi, B.: Tracking biological microorganisms in sequence of 3D holographic microscopy images. Opt. Express **15**(17), 10761–10766 (2007). https://doi.org/10.1364/OE.15.010761

124. Moon, I., Daneshpanah, M., Javidi, B., Stern, A.: Automated three-dimensional identification and tracking of micro/nanobiological organisms by computational holographic microscopy. Proc. IEEE **97**(6), 990–1010 (2009). https://doi.org/10.1109/JPROC.2009.2017563, ISSN 0018–9219

125. Schaal, F., Warber, M., Zwick, S., van der Kuip, H., Haist, T., Osten, W.: Marker-free cell discrimination by holographic optical tweezers. J. Eur. Opt. Soc. Rapid Publ. **4**(0), (2009). https://www.jeos.org/index.php/jeos_rp/article/view/09028. ISSN 1990–2573

126. Gabor, D.: A new microscopic principle. Nature **161**, 777–778 (1948)

127. Martinez-Corral, M., Saavedra, G.: The resolution challenge in 3D optical microscopy. Opt. Nanoscopy **1**(1), 68 (2009). https://doi.org/10.1186/2192-2853-1-2, ISSN 0079–6638

128. Deutsch, S.: Holographic imaging (Benton, S.A., Bove, V.M.: 2008) [Book review]. IEEE Eng. Med. Biol. Mag. **29**(3), 82–84 (2010). https://doi.org/10.1109/MEMB.2010.936543, ISSN 0739–5175

129. Goodman, J.W.: Introduction to Fourier Optics. McGraw-Hill Physical and Quantum Electronics Series. Roberts & Company (2005). http://books.google.de/books?id=ow5xs_Rtt9AC. ISBN 9780974707723

130. Ashkin, A., Dziedzic, J.M., Bjorkholm, J.E., Chu, S.: Observation of a single-beam gradient force optical trap for dielectric particles. Opt. Lett. **11**(5), 288–290 (1986). https://doi.org/10.1364/OL.11.000288

131. Svoboda, K., Block, S.M.: Biological applications of optical forces. Annu. Rev. Biophys. Biomol. Struct. **23**(1), 247–285 (1994). https://doi.org/10.1146/annurev.bb.23.060194.001335. PMID: 7919782

132. Drexler, W., Fujimoto, J.G.: Optical Coherence Tomography: Technology and Applications. Springer Science & Business Media, New York (2008)

133. Fällman, E., Axner, O.: Design for fully steerable dual-trap optical tweezers. Appl. Opt. **36**(10), 2107–2113 (1997). https://doi.org/10.1364/AO.36.002107

134. Simmons, R.M., Finer, J.T., Chu, S., Spudich, J.A.: Quantitative measurements of force and displacement using an optical trap. Biophys. J. **70**(4), 1813–1822 (1996). https://doi.org/10.1016/S0006-3495(96)79746-1, ISSN 0006–3495

135. Javidi, B., Moon, I., Yeom, S., Carapezza, E.: Three-dimensional imaging and recognition of microorganism using single-exposure on-line (SEOL) digital holography. Opt. Express **13**(12), 4492–4506 (2005). https://doi.org/10.1364/OPEX.13.004492

136. Javidi, B., Yeom, S., Moon, I., Daneshpanah, M.: Real-time automated 3D sensing, detection, and recognition of dynamic biological micro-organic events. Opt. Express **14**(9), 3806–3829 (2006). https://doi.org/10.1364/OE.14.003806

137. Rappaz, B., Charrière, F., Depeursinge, C., Magistretti, P.J., Marquet, P.: Simultaneous cell morphometry and refractive index measurement with dual-wavelength digital holographic microscopy and dye-enhanced dispersion of perfusion medium. Opt. Lett. **33**(7), 744–746 (2008)

138. Guck, J., Ananthakrishnan, R., Mahmood, H., Moon, T.J., Cunningham, C.C., Kaes, J.: The optical stretcher: a novel laser tool to micromanipulate cells. Biophys. J. **81**(2), 767–784 (2001). https://doi.org/10.1016/S0006-3495(01)75740-2, ISSN 0006–3495

139. MacDonald, M.P., Dholakia, K., Spalding, G.C.: Microfluidic sorting in an optical lattice. Nature **6965**, 421–424 (2003). https://doi.org/10.1038/nature02144

140. Buchholz, M., et al.: A multistep high-content screening approach to identify novel functionally relevant target genes in pancreatic cancer. PloS one **10**(e0122946) (2015)

141. Kemper, B., Kosmeier, S., Langehanenberg, P., von Bally, G., Bredebusch, I., Domschke, W., Schnekenburger, J.: Integral refractive index determination of living suspension cells by multifocus digital holographic phase contrast microscopy. J. Biomed. Opt. **12**(5), 054009–054009-5 (2007). https://doi.org/10.1117/1.2798639, ISBN 1083–3668

142. Moon, I., Javidi, B.: Three-dimensional identification of stem cells by computational holographic imaging. J. R. Soc. Interface **4**(13), 305–313 (2007). https://doi.org/10.1098/rsif. 2006.0175

143. Stuerwald, S., Schmitt, R.: Readjusting image sharpness by numerical parametric lenses in Forbes-representation and Halton sampling for selective refocusing in digital holographic microscopy - Errata. Version 2010. https://doi.org/10.1117/12.903693

144. Woerdemann, M., Alpmann, C., Denz, C.: Optical assembly of microparticles into highly ordered structures using Ince–Gaussian beams. Appl. Phys. Lett. **98**(11), 111101 (2011). https://doi.org/10.1063/1.3561770, ISSN 00036951

145. Rodrigo, P.J., Kelemen, L., Palima, D., Alonzo, C.A., Ormos, P., Glückstad, J.: Optical microassembly platform for constructing reconfigurable microenvironments for biomedical studies. Opt. Express **17**(8), 6578–6583 (2009). https://doi.org/10.1364/OE.17.006578

146. Matsumoto, N., Ando, T., Inoue, T., Ohtake, Y., Fukuchi, N., Hara, T.: Generation of high-quality higher-order Laguerre-Gaussian beams using liquid-crystal-on-silicon spatial light modulators. J. Opt. Soc. Am. A **25**(7), 1642–1651 (2008). https://doi.org/10.1364/JOSAA. 25.001642

147. Debailleul, M., Simon, B., Georges, V., Haeberlé, O., Lauer, V.: Holographic microscopy and diffractive microtomography of transparent samples. Meas. Sci. Technol. **19**(7), 074009 (2008). http://stacks.iop.org/0957-0233/19/i=7/a=074009

148. Choi, W.: Tomographic phase microscopy and its biological applications. 3D Res. **3**(4), 1–11 (2012). https://doi.org/10.1007/3DRes.04(2012)5

149. Thiel, M., Fischer, J., von Freymann, G., Wegener, M.: Direct laser writing of three-dimensional submicron structures using a continuous-wave laser at 532 nm. Appl. Phys. Lett. **97**(22), 221102–221102-3 (2010). https://doi.org/10.1063/1.3521464, ISSN 0003–6951

150. Gittard, S.D., Nguyen, A., Obata, K., Koroleva, A., Narayan, R.J., Chichkov, B.N.: Fabrication of microscale medical devices by two-photon polymerization with multiple foci via a spatial light modulator. Biomed. Opt. Express **2**(11), 3167–3178 (2011). https://doi.org/10.1364/ BOE.2.003167

151. Soukoulis, C.M., Wegener, M.: Past achievements and future challenges in the development of three-dimensional photonic metamaterials. Nat Photon **5**(9), 523–530 (2011). https://doi. org/10.1038/nphoton.2011.154, ISSN 1749–4885

152. Kafri, O., Kimel, S.: Theory of two-photon absorption and emission second-order saturation effect. Chem. Phys. **5**(3), 488–493 (1974). https://doi.org/10.1016/0301-0104(74)85052-4, ISSN 0301–0104

153. Encyclopedia of Nanotechnology. Springer, Dordrecht (2012) (Springer reference). ISBN 9048197511

154. del Campo, A., Greiner, C.: SU-8: a photoresist for high-aspect-ratio and 3D submicron lithography. J. Micromech. Microeng. **17**(6), R81–R95 (2007). https://doi.org/10.1088/0960-1317/17/6/R01, ISSN 0960–1317

155. Roxworthy, B., Ko, K.: Application of plasmonic bowtie nanoantenna arrays for optical trapping, stacking, and sorting. Nano Lett. **12**, 796–801 (2012)

156. Donner, J.S., Baffou, G., McCloskey, D., Quidant, R.: Plasmon-assisted optofluidics. ACS Nano **5**(7), 5457–5462 (2011). https://doi.org/10.1021/nn200590u, ISSN 1936–0851

157. Fischer, U.C.: Submicroscopic pattern replication with visible light. J. Vac. Sci. Technol. **19**(4), 881 (1981). https://doi.org/10.1116/1.571227, ISSN 00225355

158. Kandulski, W.: Shadow nanosphere lithography. Ph.D. thesis. Universität Bonn (2007)

159. Fischer, U.C.: Submicroscopic pattern replication with visible light. J. Vac. Sci. Technol. (1981). https://doi.org/10.1116/1.571227

160. Ke, M.-Y., Wang, C.-Y., Chen, L.-Y., Chen, H.-H., Chiang, H.-L., Cheng, Y.-W., Hsieh, M.-Y., Chen, C.-P., Huang, J.: Application of nanosphere lithography to LED surface texturing and to the fabrication of nanorod LED arrays. IEEE J. Sel. Top. Quantum Electron. **15**(4), 1242–1249 (2009). https://doi.org/10.1109/JSTQE.2009.2016433, ISSN 1077–260X

161. Zhang, X., Yonzon, C.R., Van Duyne, R.P.: Nanosphere lithography fabricated plasmonic materials and their applications. J. Mater. Res. **21**(5), 1083–1092 (2006). https://doi.org/10.1557/jmr.2006.0136, ISSN 2044–5326
162. Hoffmann, J.M., Yin, X., Richter, J., Hartung, A., Mass, T.W.W., Taubner, T.: Low-cost infrared resonant structures for surface-enhanced infrared absorption spectroscopy in the fingerprint region from 3 to 13 μm. J. Phys. Chem. C **117**(21), 11311–11316 (2013). https://doi.org/10.1021/jp402383h
163. Righini, M., Volpe, G., Girard, C., Petrov, D., Quidant, R.: Surface plasmon optical tweezers: tunable optical manipulation in the Femtonewton range. Phys. Rev. Lett. **100**, 186804 (2008). https://doi.org/10.1103/PhysRevLett.100.186804

Chapter 5
Summary on DHM Methods

In this work, the combination of optical methods, mainly based on holographic principles for marker-free imaging, real-time trapping, identification and tracking of micro objects is investigated. Therefore, first an overview of Digital Holographic Microscopy (DHM) and holographic optical tweezers as well as non diffracting beam types is given for minimal-invasive, real-time and marker free imaging of micro and nano objects.

In the experimental part of this work, first advanced reconstruction methods adapted to broad spectrum light sources are proposed, investigated and demonstrated on technical surfaces and biologic specimen with different coherent (NdYAG, HeNe-laser) and low coherent light sources (LEDs based on InGaN, GaAsP and AlGaInP, SLDs, supercontinuum light source). The experimental results show, that low coherent light sources - including LEDs with coherence lengths of 4–16 μm - offer reduced phase noise in comparison to lasers of up to 63% for spatial and 55% for temporal phase shifting techniques. Drawbacks of LEDs are the sensitivity of alignment and relatively high exposure times of up to $T_B = 0, 5$ s, since depending on the alignment less than 4% ($\approx(360 \pm 20)$ nW) of the light power can be exploited after Fourier filtering with a pinhole of 25 μm diameter.

Since a reliable region selective numerical readjustment of the focus of the complex wave fronts is of particular interest in DHM especially for compensation of displacements and for tracking in long-term investigations of living cells, approaches for an improved subsequent numerical refocusing functionality has therefore been investigated. In order to enhance this time consuming method of numerical searching of the best focus in every sub-area of a field of view, a Halton point set with low discrepancy has been implemented. his method avoids a narrowing of the field of view leading to a loss of information around the focus plane by blurring. For the concept of numerical parametric lenses for correcting aberrations in the reconstructed wave front caused by the setup, the polynomial basis by Forbes is implemented and adapted for the needs of DHM for reduction of the number of parameters for these

© Springer Nature Switzerland AG 2018
S. Stuerwald, *Digital Holographic Methods*, Springer Series
in Optical Sciences 221, https://doi.org/10.1007/978-3-030-00169-8_5

parametric lenses and for facilitating the handling of this key feature of DHM for users.

For holographic optical tweezers (HOTs), design concepts for the optical layout are investigated and optimised with optical simulations and experimental methods. The system module for adding HOTs to a commercial microscope basis is characterised with regard to stability and diffraction efficiency. The functionality of the combined setup is demonstrated in experiments with micro-fluidic environments and can be performed with stably arranged or distanced objects and cells, which allows to examine e.g. their deformations, changes in volume or reactions to certain artificially manipulated surface types.

Because of their possible functionality for the arrangement of micro-particles and structures, several types of non-diffracting beams are discussed as well as their application for optical manipulation. It is demonstrated, that holographic optical tweezers not only open up the possibility of generating multiple dynamic traps for micro and nano particles with forces in the pico and nano newton range, but also the opportunity to exert optical torque with special beam configurations like Bessel beams, which can facilitate the movement and rotation of particles also by generating micro-fluidic flows. Moreover, the utilisation of further non diffracting beams like Matthieu beams and combined parabolic and Airy beams as stable beam types for arranging particles and their distances in specific geometries are investigated. Matthieu beams as an example for relatively complex non-diffracting beams are demonstrated with emphasis on innovative schemes for optical trapping.

The experimental realisation of these extraordinary laser beams, which may offer so called "self healing" characteristics, is also performed with a spatial light modulator based on liquid crystals (LCoS-type).

The possibility of reducing the trapping light power and therefore increasing the trapping efficiency is basically shown on arrays of $2\frac{1}{2}$-dimensional gold nano antennas on glass substrate with the form of triangles (also known as bowtie antennas), which were fabricated with nano-sphere lithography. Here, the local field enhancements and reflection properties of the modified glass substrate lead to verifiable improvement of 12% lower trapping light power in case of the utilised 1 μm diameter micro beads (partially transparent spheres).

In a last part of this work, the functionality of direct laser writing based on a two photon absorption process in a negative photoresists with a continuous wave laser has been investigated with a slightly modified DHM-HOT-system. Therefore, the optics have been adapted to 532 nm and the system has been re-optimised. By implementation of software based slicing-algorithms for a data processing of the CAD-data for the writing process and by automating the trajectories of several multi-focal dynamic traps, a multi spot writing functionality has been demonstrated. The available laser power of 200 mW allowed two writing spots simultaneously at maximum with a laser power of each spot in the range of 40 ± 8 mW, since the losses in the setup are relatively high mainly due to the partial homogenisation of the beam intensity distribution. Here, the available laser power showed to be only sufficient in case of application of Bessel beams of first order, ensuring that the intensity threshold of the nonlinear absorption process is exceeded. In case of one writing spot, the Bessel beam was not needed. The detection of the interface of glass substrate and photoresist, where a writing process has to start in order to bond a structure to the surface, is performed by recording the correlation function of an SLD or LED. The added feature of this multifunctional setup is demonstrated with written line patterns in SU-8 photoresist.

Currently, to my knowledge no other system is existing that offers holographic optical tweezers, simultaneous digital holographic imaging and multi-focal direct laser writing additionally to the possibility of bright field and fluorescence laser scanning microscopy in one setup. A system with the investigated range of functions provides unprecedented possibilities to cell analysis, handling and manipulation of micro and nano objects.

Chapter 6
Prospects

In this work a system combination of digital holographic microscopy, holographic optical tweezers and a laser scanning fluorescence microscope is developed and furthermore applied for dynamic and multi-focal direct laser writing in photoresists on a 2-photon polymerisation basis - also known as 3D-lithography - which shows potential for a variety of further development steps. Above all the research of utilizing a femto-second laser to increase the efficiency of the 2PP process would be wise. It is expected that the optical power can be significantly reduced in comparison to a CW-laser, increasing the repeatability and speed of the writing process. The application of ultra-short laser pulses however limits the applicability of the current spatial light modulator since the temporal phase stability would not be sufficient for this laser radiation. Here alternative models (Hamamatsu, Boulder Optics) have to be used, which also enable a higher rate of repetition, but require - like the femto-second laser - higher system investments. For a possible permanent use of the system for the three dimensional direct laser writing, an automated search for boundary surfaces between substrate and photoresist on the basis of the acquired short coherent correlogram is advisable to decrease the set-up time and increase the reliability of the system. Additionally the implementation of an integrated graphical user interface (GUI) that can control all functionalities of the modular system and also slice imported 3D-Data would be helpful. In regard to an eventual online deployment of the methods developed in this work, a runtime optimisation of the implemented algorithms is of practical importance. For the writing of three dimensional micro and nano structures with negligible *stitching losses* on large-scale structures of areas $\geq 300 \times 300 \, \mu m^2$, a high precision xyz-positioning unit for the main microscope system would be needed for permitting a higher positioning accuracy.

Furthermore the extension of the metrological combination possibilities is possible since additional optical paths can be easily integrated into the already existing one using dichroic mirrors and beam splitters. The most important point to consider when feeding additional beam paths into the microscope is the chromatic aberration for the microscope objective and the wavelength specific blooming of the optical

© Springer Nature Switzerland AG 2018

S. Stuerwald, *Digital Holographic Methods*, Springer Series
in Optical Sciences 221, https://doi.org/10.1007/978-3-030-00169-8_6

elements. Different structured illumination methods [1] as well as "Stimulated Emission Depletion"-microscopy (STED, [2]) for imaging and structuring methods are possible here, which has been partly employed once for 3D-Laserwriting [3]. The addition of a beam path for a confocal scanning microscope mode as well as an integrated RAMAN-spectroscopy are theoretically possible. The higher complexity of the overall system, however, would demand a high-grade optimisation of every subsystem which is not possible with the commonly shared microscope optics in the utilised setup.

The used Linnik-interference microscope module facilitates a higher sensitivity on transparent phase objectives. However, for an optimised geometry and illumination further investigations are advisable. For the transparent, cellular samples in aqueous solution that were investigated using the Linnik-setup, one could additionally try to use the air-water boundary as an alternative to the mirror. This would significantly simplify the sample preparation during a measurement.

In microscopy, halogen light sources are still used most for object illumination because of their high luminous intensity needed especially for magnifications $>50\times$ despite their high spectral fraction in the near infrared wavelength area which causes undesired warmth and thermal drifts. Since the light power of LEDs is still under intensive development, the prospects for increased application of LEDs are promising. This would allow a simultaneous, low cost utilisation of a phase shifting in various microscope setups and thus has the potential to extend the functionality of many standard setups in the future. Especially the low acquiring costs and the continuing development of LEDs with higher light power suggest a further investigation of new LEDs.

A first approach towards realisation of an inexpensive and portable microscope has been recently realised by application of lens-less digital holography in order to omit lenses and improve miniaturisation [4]. This setup does not allow quantitative phase contrast microscopy yet, but shows the potential of this technology for future developments and applications since adding more functionality to such setups is considered to be feasible.

A relatively innovative approach for digital holographic microscopy would be the application of the 3D-light field technology based on plenoptic cameras which recently became commercially available in a sufficient imaging quality (Raytrix) and that allows to determine the direction of the light for every pixel on a camera with the help of microlens arrays [5–7]. This reduces the effective resolution of an image sensor by typically a factor of 4, but in case of an oversampling the effective resolution of the system would not be affected. The main advantage of such a light field camera would consist in an increased depth-of-field by a factor of $\approx$6 for standard cameras that would enhance the refocussing capability of the developed setup significantly since the software based refocusing capability of the digital holographic system could then be extended by that of the light field camera. Till now, these cameras have not been investigated for microscopy.

References

1. Gustafsson, M.G., Shao, L., Carlton, P.M., Wang, C.J.R., Golubovskaya, I.N., Cande, W.Z., Agard, D.A., Sedat, J.W.: Three-dimensional resolution doubling in wide-field fluorescence microscopy by structured illumination. Biophys. J. **94**(12), 4957–4970 (2008). https://doi.org/10.1529/biophysj.107.120345, ISSN 0006–3495
2. Rittweger, E., Han, K.Y., Irvine, S.E., Eggeling, C., Hell, S.W.: STED microscopy reveals crystal colour centres with nanometric resolution. Nat. Photonics **3**(3), 144–147 (2009). https://doi.org/10.1038/nphoton.2009.2, ISSN 1749–4885
3. Soukoulis, C.M., Wegener, M.: Past achievements and future challenges in the development of three-dimensional photonic metamaterials. Nat. Photonics **5**(9), 523–530 (2011). https://doi.org/10.1038/nphoton.2011.154, ISSN 1749–4885
4. Lee, M., Yaglidere, O., Ozcan, A.: Field-portable reflection and transmission microscopy based on lensless holography. Biomed. Opt. Express **2**(9), 2721–2730 (2011). https://doi.org/10.1364/BOE.2.002721
5. Lippmann, G.: Epreuves reversibles donnant la sensation du relief. Journal de Physique Theorique et Appliquee **7**(1), 821 (1908)
6. Adelson, E.D., Bergen, J.R.: The plenoptic function and the elements of early vision. In: Landy, M.S., Movshon, J. A. (eds.) Computational Models of Visual Processing. MIT Press, Cambridge (1991). ISBN 9780262121552
7. Ng, R. et al.: Light field photography with a hand-held plenoptic camera. In: Stanford Tech Report CTSR 2005-02, University of Stanford (2005)

Appendix A

A.1 Developed Software

A.1.1 Implemented Software for DHM

HoloControl

The software interface "HoloControl" (Holographic system Control, data processing schematic in Fig. A.1 data processing schematic in Fig. A.2) allows an automated recording of digital holograms, where the integrated control of the long travel rangepiezo phase shifter allows the application of simultaneous spatial and (or) temporal phase shifting methods. In case of low coherent light sources, the estimated coherence length of the source is required in order to determine the effective coherence length automatically by recording the cross correlation function in an efficient initialisation routine before start of measurements. In case the coherence length under-runs a critical threshold, only temporal phase shifting is permitted by the software. The phase shifted holograms are recorded by the image capturing device and stored on a computer. A log-file ensures recording of the settings of each measurement or sequence of a time lapse. The system control for the movement of the NMM-1 nano measuring and positioning system is optionally included. A specific option in combination with the contrast algorithms allows to find an interface (e.g. of glass and liquid) automatically which is helpful for the SLM-based 2PP-writing option for direct laser writing. For measuring topographies in refection mode, a stitching routine is implemented which works for relatively plane objects.

HoloCalc

The graphical user interface "HoloCalc" (Hologram Calculation, Fig. A.3) represents a user friendly software interface for reconstruction of temporal and *or* spatial phase shifted holograms. The algorithm for the reconstruction method can be chosen as well as automated filter and autofocusing routines for auto detection of the sharpest image plane. Furthermore, an algorithm for 3-dimensional tracking of chosen objects is implemented. The results are stored as complex matrices or as separated phase and

© Springer Nature Switzerland AG 2018
S. Stuerwald, *Digital Holographic Methods*, Springer Series
in Optical Sciences 221, https://doi.org/10.1007/978-3-030-00169-8

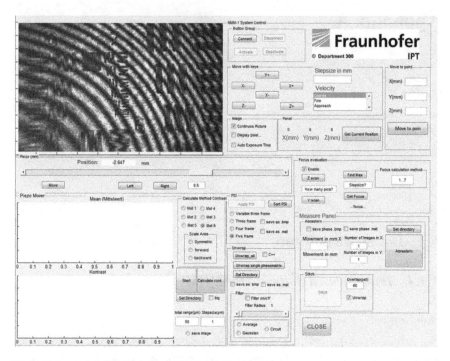

Fig. A.1 Layout of GUI "HoloControl"

amplitude information in an optional *hdf5* data format. A schematic flowchart for the computational procedure is depicted in Fig. A.3. For a continuous reconstruction of recorded holograms of a camera in a microscope setup, a real time option has been implemented (Fig. A.4). According to the performance of the computer and the amount of continuously adapted reconstruction parameters, the display of the quantitative phase contrast image lasts approximately 2 s for a standard computer (dual core, 3 GHz).

HoloAnalyzer

The GUI "HoloAnalyzer" (Hologram Analyzer, Fig. A.5) permits an illustration and quantitative analysis of reconstructed holograms which have been stored by the GUI "HoloControl" in a binary, lossless format. This allows a faster handling and analysis of the results especially of time lapses with a big amount of data, whose repeated reconstruction would require extensive calculation power.

A.1.2 Implemented Software for HOTs

The algorithm for the realisation of phase patterns in the Fourier plane of a microscope setup is explained in Sect. 2.6.5. The implemented software for dynamic hologram computation is based on LabVIEW and an output through a graphics card port (DVI).

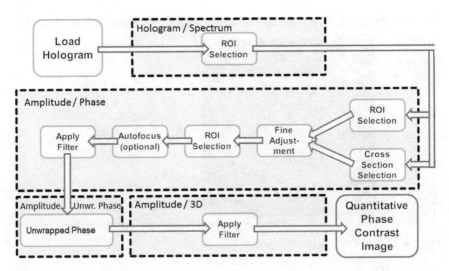

Fig. A.2 Schematic diagram of data processing in implemented software "HoloCalc"

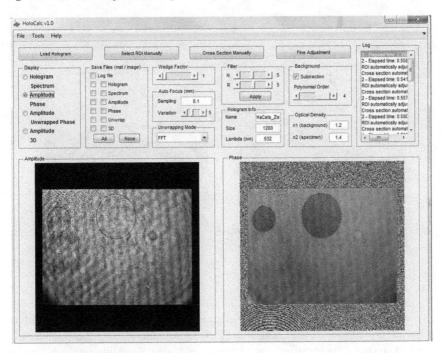

Fig. A.3 Layout of GUI "HoloCalc"

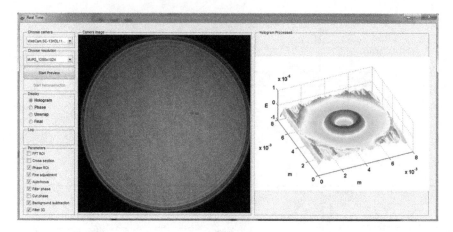

Fig. A.4 Layout of the real time option in GUI "HoloCalc"

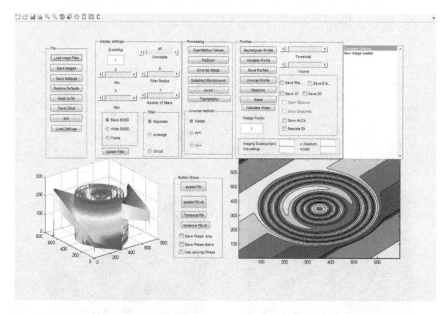

Fig. A.5 Layout of GUI "HoloAnalyzer"

Two versions of the software were implemented - one entirely based on CPU calculation and another where the hologram calculation is outsourced to the graphic card. The CPU based software offers the advantage of higher flexibility on the hologram computation. For example, various calculation algorithms can be selected. Furthermore, system aberrations can be already corrected in the software by calculating a correction image based on the definition of the Zernike coefficients Sect. B.3.8 and superimposing it directly with the trap hologram. However, the full hologram calculation is carried out in the LabVIEW program, whereby the CPU of a standard

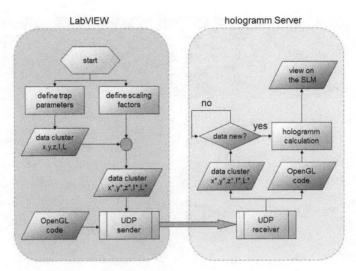

Fig. A.6 Data flow of the application software

office PC is fully loaded even with a low number of traps and it is no longer possible running the program in real time. Additionally, the program is less flexible in the trap positioning than the GPU-based version.

The GPU based version offers little influence on the hologram computation and algorithm. With a Matlab or LabView-based GUI the spot coordinates are passed to the GPU for the final hologram calculation. For this, another program is needed (server executable), which performs the hologram calculation using OpenGL on the graphics card (required: shader model 4 or higher). LabVIEW communicates with this program through the User Datagram Protocol (UDP) interface. The hologram computation on the graphics card is much more efficient than the one on the CPU, that is why the hologram computation of many trap positions runs faster (≈ 100 Hz) than the CPU-version (≈ 10 Hz). Also the production and positioning of traps is much more flexible in this version.

The schematic program sequence is shown in Fig. A.6. After starting the software trap parameters can be set. These are the spatial coordinates (x, y and z), the intensity (I) and the azimuthal index (L). In addition to manual input, the x and y-coordinates can be dynamically changed by moving the point with the mouse. The five parameters of all traps are summarised in a data cluster. Furthermore, scaling factors have to be set,which are taking into account the size of the SLM and the mapping of this on the object plane. Scaled by these values, the new data cluster is passed to the SubVI "UDP emitter" which includes the code string that compiles the shader program responsible for the hologram rendering. The SubVI establishes a UDP connection via a port of the computer that is read out from the hologram server program. The hologram server executes the OpenGL code with the data from the data cluster, and renders the hologram on the graphics card. The final rendered hologram is then

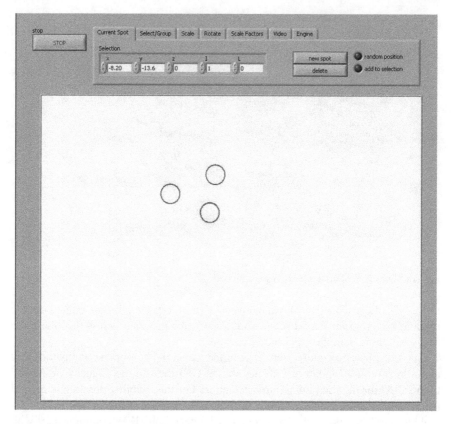

Fig. A.7 LabVIEW front panel

generally sent to the SLM panel via the second DVI output of the graphics card. Once new data is available in the data clusters, a new hologram rendering is performed.

The user interface of the software is shown in Fig. A.7. By displacement of the circles in the image, the lateral coordinates can be changed. The five parameters of the selected trap can also be entered manually in the line above the screen (field of view). In the other program it is possible to group traps together or rotate them around a specific point. A change of the scaling parameter is also possible. A schematic layout of the position detection program for optimised, partial automated trapping is depicted in Fig. A.8.

System-related parameters like the wavelength of the used laser and the effective focal length of the microscope objective are included in the calculation of the hologram. To avoid errors in the optical reconstruction, these must be taken into account when calculating the hologram. The software provides the option to set the wavelength of the laser and the focal length of the selected microscope objective in the engine section. Furthermore, the dimensions of the SLM area projected into the microscope entrance pupil are entered. This is calculated as the product of the

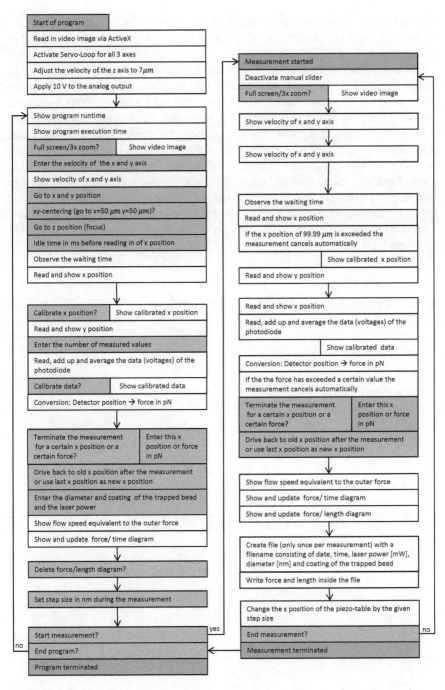

Fig. A.8 Schematic layout of position detection and force measurement software for trapping

diameter of the SLM inner circle and the figure factor of the system: 1080 Pixel ·
8 μm/Pixel · 1,3333 ≈ 11520 μm. The entered system parameters are considered in
the calculation of the hologram.

The controlling of the used CCD camera (Retiga-2000RV of QImaging, data in
Appendix B.1) was integrated into the LabVIEW program. The camera manufacturer
offers a LabVIEW SubVI that can control and read out the camera. This SubVI has
been implemented in the video tab of the trap software. For the exposure time a
control element was integrated. The camera image is displayed on the front panel
below the trap spot circles. A calibration of the spot position of the trap software
and of the camera image is performed by scaling the image size. To align the x and
y coordinates of the camera and trap software, the x and y coordinates of the trap
software have to be reversed. This is realised in the implemented SubVI "*gratings and
lenses.vi*" by swapping the linking of x and y before combining them into a shared
array. Due to the complexity an illustration of the program structure (the so-called
back panels) is hardly possible in clearly arranged way.

A.1.3   OpenGL Code for Hologram Rendering

```
<data>
<shader_source>
uniform float k;
uniform float f;
uniform vec2 slmsize;
uniform vec4 spots[50]; //maximum 25 spots
uniform int n;

void main(){
// basic gratings and lenses for a single spot
  vec2 uv = (gl_TexCoord[0].xy-0.5)*slmsize;
  vec4 pos = vec4(k*uv/f, k*dot(uv,uv)/(2.0*f*f),atan(uv.x, uv.y));

  float phase, real=0.0, imag=0.0;
  for(int i=0; i<2*n; i+=2){
    phase = dot(pos, spots[i]);
    real += spots[i+1][0] * sin(phase);
    imag += spots[i+1][0] * cos(phase);
  }
    phase = atan(real, imag);
    float g = phase / 6.28 + 0.5; // map -pi to pi onto zero to one

    gl_FragColor=vec4(g,g,g,1.0);

//   gl_FragColor = vec4(1.0,0.0,1.0,1.0);
}
</shader_source>
</data>
```

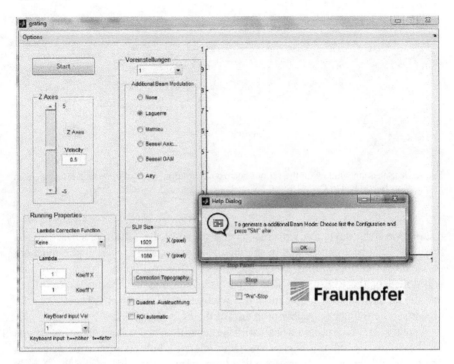

Fig. A.9 Matab GUI

A.1.4 Graphical User Interface for Complex Beam Configurations

The implemented GUI comprises the complete theoretical groundwork given in Sect. 2.9. Although there exist several software approaches for controlling the laser traps, however, with this solution the combination of the control of multiple spots and the production of various beam configurations is fulfilled in one program. Since the generation and control of multiple spots is not outsourced to the graphic card, this results in a loss of speed.

Program Structure

The structure of the program "Gratings" schematically depicted in Fig. A.11.

When the start of the program is called, the following interface appears (see Fig. A.9). First, multiple pre-sets can be made by the user. These include the number of traps, the additional beam configuration, SLM- properties, hologram resolution and automatic search for the region of interest (ROI).

The selection of the number of traps is done using a predefined Popup-menu. For the choice of the beam configuration a symbol must be activated, resulting in a further window for selection of specified beam properties (see Fig. A.10). Furthermore, the SLM resolution and/or a topographic correction can be selected. In the correction

Fig. A.10 Selection window for the various beam configurations with additional parameters (from left to right: Bessel beam: OAM, Bessel beam: axicon, Laguerre, Mathieu, Airy)

routine the form deviation with regard to a plane of the reflective surface of the modulator can be compensated by adding the corresponding inverted phase pattern of the measured topography (Sect. A.2.3). The other two options allow automatic adjustment of the ROI to the size of the user interface and the utilisation of the modulator in a square format instead a widescreen format which accelerates the process and eliminates possible problems of the PC's resources.

Another part of the program deals with the generation and storage of the holograms for the different number of spots. As for each spot a hologram must be generated and saved, this requires a corresponding high memory capacity of the PC (>6Gb recommended).

After selection of the main options the selected settings are loaded and a video signal of the microscope camera is displayed. Trapping and moving can be done in two ways using the keyboard or mouse. In this case, if several traps are present, the current (last) spot is controlled. For the keyboard controls a speed (1–50 (Pixels)/(actuation)) can be selected. Depending on the number of traps these are shown by different symbols (star, circle, rhombus, etc...). For a precise function of the program, some parameters of the algorithm as a nonlinear inclusion of the grating-constant, with four suggested functions, and the constant itself can be adjusted. In addition to the lateral displacement of traps (realised by phase gratings) the axial manipulation by phase lenses is implemented. For this purpose the z-slider can be moved in the left part of the GUI. The associated function comprises the second part of the *Grating and Lenses* algorithm. A control with help of the keyboard is also implemented for this feature. The adjustment is made with the keys *h* for the rising and *t* for lowering case. The increment of this can be set next to the z-slider. Overall, the traps can be moved individually and independently both in axial and in radial direction. By pressing the stop panels the calculation algorithm is interrupted and the hologram is closed (according to Fig. A.11).

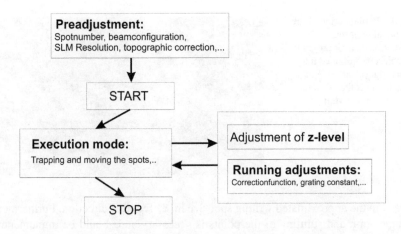

Fig. A.11 The schematic representation of the program created

A.1.5 Extension of HOT-Software for Multifocal Direct Laser Writing

In order to successfully employ multi-focal laser writing, a few features have to be integrated into the optical tweezers control software (see Fig. A.13). The most important part is the addition of a read-in routine that accepts the *csv* files from the slicer and then moves the given number of spots through the given positions. This routine reads in the file and automatically moves as many spots as coordinates are supplied in the same column.

In the following, first the slicer functionality is explained for pre-calculation of the required data format for the direct writing routine.

Slicer

An STL file is a triangular mesh of the complete three dimensional object. This however can not be fed into the optical tweezers software, since optical tweezers only need single spots or intensity distributions. Therefore, a slicer software is necessary, in order to translate a STL file into a format where it can be read in and used easily by the optical tweezers software.

In contrast to the previously mentioned Nanoscribe system, the goal of this work is to write with a SLM. Nanoscribe, however, writes with a fixed laser spot while the stage holding the sample is moved. Such a setup suggests a slicer which transforms the mesh into curves or lines that are subsequently written. In this case, due to the use of an SLM, a pixelated approach is more intuitive. The SLM can display a holographic pattern and change it 60 times per second. This results in a pixelated behaviour of the writing laser spots, which are moved by an increment every 1/60 s. The goal for the slicer software is therefore to take an arbitrary (closed) triangular mesh and calculate a point cloud that represents that object. This has to be done in

Fig. A.12 Illustration of the functionality of the implemented slicer program: An object gets divided into layers (right), and each of those layers is then further divided into rays (left)

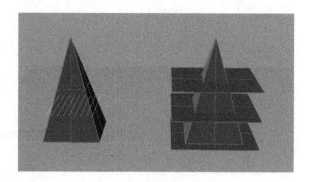

respect to the approximated writing spot size in X, Y and Z direction. Furthermore, a numeration and splitting of the points is necessary and should be implemented. A numeration is intuitive since a three dimensional structure can not be written from the top. The software has to ensure that the lowest point is written first, and the highest point is written last. Otherwise the written structure would float away in the polymerisation liquid. After numeration the points can then be splitted and distributed onto the writing spots in order to achieve a multi-focal process.

To achieve this goal, a bounding box around the triangular mesh is generated first. The maximal and minimal coordinate values of all triangles are taken, thus creating a box which can be extended by a user controllable safety margin. Afterwards a grid is constructed filling this bounding box in respect to the voxel size which can be modified. This process is independent in all three dimensions, therefore an asymmetrical spot can also be used. This grid is then divided into z-layers which are subsequently divided into rays (see Fig. A.12). Every ray is then intersected with every triangle to determine which parts of the ray are inside and which are outside of the structure. An even number of intersections means outside and an odd number means inside. For the ray-triangle intersection an algorithm of Möller and Trumbore [1] is employed. This algorithm is especially memory saving and also calculates the barycentric coordinates and therefore the distance from the origin of the ray to a possible intersection with a triangle.

In this approach all triangles are intersected with all rays. Although this method is simple and time consuming, it can easily be computed simultaneously. On modern PCs which often feature multi-core processors, such parallelisation capability is important and saves calculation time. Even with the speed advantage through parallel computation, this is, however, not the fastest way to solve the problem. For example, one possible approach would be to sort the triangles into different areas, and when computing the intersections of a ray only this subset of triangles would have to be taken into consideration. Since this work is focused on the laser writing process itself, the software is not optimised beyond parallel execution.

Having intersected every triangle with every ray, one knows now where the rays hit the surface and therefore which parts are inside and which are outside of the structure. This knowledge can subsequently be used to break up the rays into points and write all points that are inside into an array. The points are then numbered according to

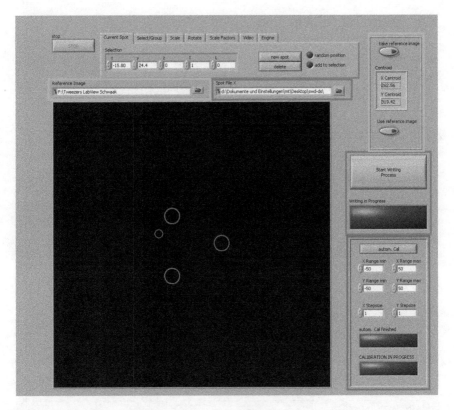

Fig. A.13 Extended optical tweezers GUI - with added direct laser writing elements (red), the zero-image subtraction (green), and the automatic calibration (blue)

the Z and then X and Y value. For every Z-Value the available points are counted and distributed onto the writing spots so that the output of the slicer program is a list of coordinates for a choosable amount of writing spots. This list is saved as three comma separate value (csv) files: one for the x-, one for the y- and one for the z-value of all spots.

Although an own slicing software functionality has been developed, the benefit of this application for full 3D-structures could not be shown due to the insufficient microscope stage accuracy. The HOT-software has been extended for this new application of direct laser writing (see also description in Appendix A.1.5).

The last software sided implementation is a calibration routine. As seen in Fig. A.13 one can adjust the range and step-size of the x and y coordinates. The system then automatically scans the area with one spot and evaluates the centroid of the acquired image. With the given spot position and the measured spot position, a polynomial fit can then be performed with help of the matlab toolbox 'polyfitn'. This polynomial calculation and correction is then implemented in the control software.

A.2   Technical Information

In the following a brief overview of several important technical data of the used hardware as well as the programmed software is given.

A.2.1   Optic Simulation: System Data of HOT-Module

Oberfl: Typ		Kommentar	Radius	Dicke		Glas	Halber Durchm..	
OBJ	Standard		Unendlich	Unendlich			0.000	
STO	Standard		Unendlich	10.000			1.750	
2*	Standard	ACN254-050-B	-27.050	2.000		N-BAF10	12.700	U
3*	Standard		33.910	4.000		N-SF6HT	12.700	U
4*	Standard		-1330.500	140.928	V		12.700	U
5*	Standard		-2000.000	4.000		SF10	12.700	U
6*	Standard		96.610	4.000		N-LAK22	12.700	U
7*	Standard	AC254-200-B	-106.410	40.000			12.700	U
8	Standard	Polarisator	Unendlich	181.100			8.994	
9	Koordinat..			0.000		–	0.000	
10*	Standard	SLM	Unendlich	0.000		MIRROR	4.320	U
11	Koordinat..			-30.000		–	0.000	
12	Standard	Analysator	Unendlich	-243.870			8.994	
13*	Standard	AC254-150-B	-83.600	-4.000		N-LAK22	12.700	U
14*	Standard		89.330	-3.500		N-SF6HT	12.700	U
15*	Standard		1330.500	-155.630	V		12.700	U
16	Koordinat..			0.000		–	0.000	
17	Standard	Dichroitische..	Unendlich	0.000		MIRROR	2.766	
18	Koordinat..			58.000		–	0.000	
19	Standard	Port	Unendlich	131.000			5.240	
20	Paraxial	Tubuslinse		120.000			11.390	
21	Paraxial	MO		3.200			11.390	
IMA	Standard		Unendlich	–			4.086E-014	

A.2.2   Callibration Using a Wavefront Sensor

In response to the incident beam, the focal points are generated by the microlens array on the CCD sensor array at different positions and an analysis on the intensity, Zernike aberrations and the wave front form can be performed. For this purpose the sensor is placed at the desired location and the corresponding program can be started. For the analysis many options like the Wave Front, Spot Field, Zernike Coefficients, etc. can be utilised. In this work the Zernike coefficients were mainly utilised because of their quantitative character.

For an error-free operation of the imaging and trapping system at many positions a plane wave front is required since a uniform intensity distribution has to be assumed for the realisation of the desired beam configuration (Sect. 2.9.1). For that the Zernike coefficients of the 3rd order (Sect. B.3.8) are mainly used, which allow the measure-

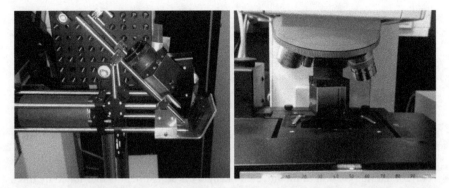

Fig. A.14 Photo: WFS before the SLM (left) and instead of the lens (right)

Fig. A.15 The mounting and adjustment possibilities of the spatial light modulator (left) and the dichroic mirror (right)

ment of the curvature of the wavefronts. The adjustment can be done at two places of the setup. First of all the wave front before the SLM has to be investigated to guarantee the correct orientation of the fibre coupler and the correct position of the two lenses (L1 and L2, Fig. 4.31).

Furthermore it is essential to illuminate most of the area of the modulator and to find a compromise with regard to the loss of power by the expansion. After the correct illumination of the SLM, the wave front sensor has to be placed in the mount of the microscope objective (Fig. A.14 (right)). For this purpose the Zernike coefficients have to be investigated as well. The adjustment is mainly performed on three points of the setup. First the reflected beam from the SLM has to be positioned parallel to the beam direction before the third lens (L3) (Fig. 4.31). This can be done by the correct tilt of the SLM which can be varied on the back side of the modulator by adjustment of three screws (see Fig. A.15 and the entire setup in Fig. A.17).

Secondly the expansion of the reflected beam has to be changed to couple the complete wave front into the microscope objective correctly. For this purpose the z-position of the third lens has to be varied. Finally a dichroic mirror position ensures

Fig. A.16 Various analysis features for calibration of the wave front of the trapping laser: Wave front (left), Field Spot (center), Zernike Coefficients (right)

a correct lateral position of the injected beam. This adjustment is facilitated by three screws at the side of the mirror (see Fig. A.15).

During the entire adjustment the Zernike coefficients are the most important indicators for the investigation and have to be minimised. However for the proper functionality of the *Grating and Lenses* algorithm the wave front has not only to be plane, but requires exact centring and zero tilt. Thus the investigation of the second and third Zernike coefficients (Sect. B.3.8) are performed with the WFS at the place of the microscope objective. In the following a few screen-shots of the WFS-application with the characteristic are shown (Fig. A.16).

A.2.3   Interferometrically Determined form Error of SLM

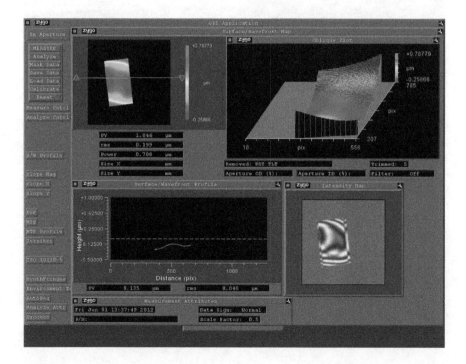

A.2.4 Photos of Module for Optical Trapping

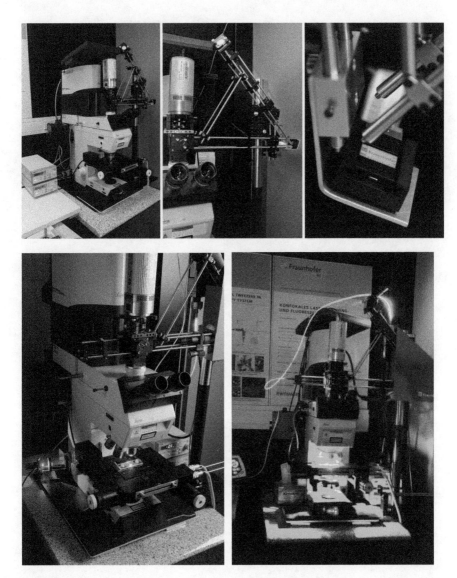

Fig. A.17 DHM-HOT setup in operation (left) and with direct laser writing mode with a 532 nm laser (right)

Appendix B

B.1 Specifications of Utilised Systems

B.1.1 Super Continuum Light Source

The utilised super continuum broad band light source Super K Extreme (NKT-Photonics) is a photonic crystal fibre laser offering a dense light spectrum in the range of [460, 2400] nm. The maximum total light power constitutes ≈ 1.7 W with a variable repetition rate of 1–80 MHz and a variable pulse suppression ratio of $>18\,000$. The crystal fibre is pumped with a pulse of 5 ps duration which generates an output pulse with the length of 500–700 ps. The light source is primarily intended for fluorescence microscopy with multiple markers. Further applications are the microscopy modes *FLIM*, *FRET*, *Diffuse Optical Tomography* and *TCSPC*.

The selection of the wavelength range is performed with the help of acousto optical tunable filters (AOTF), which allow a wavelength tuning over a specified range (VIS/nIR, VIS/IR, nIR/IR). Each AOTF permits 8 simultaneous, freely tunable wavelength channels.

Coheras
See Table B.1 and Fig. B.1.

Trap Laser I ($\lambda = 975$ nm)
See Table B.2.

Trap Laser II ($\lambda = 1064$ nm) **Phase Contrast Laser** ($\lambda = 532$ nm)
See Tables B.3 and B.4.

Wave Front Sensor
See Table B.5.

© Springer Nature Switzerland AG 2018
S. Stuerwald, *Digital Holographic Methods*, Springer Series
in Optical Sciences 221, https://doi.org/10.1007/978-3-030-00169-8

Table B.1 Technical specifications of super continuum light source *SuperK Extreme* from *NKT-Photonics*

Wavelength range	460–2400 nm	
Total light power	≈1.7 Watt	
Master seed laser repetion rate	80 or 40 MHz	
Suppression ratio	>1 : 8000	
Operation mode	Constant pulse energy	
Delay shift betw. pulse picker ratio	<±250 ps	
AOTF	Type: **VIS**	Type: **VIS/nIR**
Wavelength range	450–750 nm	640–1100 nm
Filter bandwidth in nm	≈3.5(@480)–7(@650)	≈2(@640)–5(@1100)
Number of tunable lines	1–8	1–8

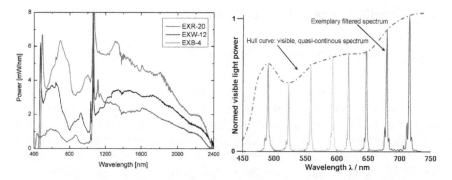

Fig. B.1 Left: Complete spectrum of the supercontinuum light source Super K Extreme. The measurement follows a special method to allow adequate scaling within the different measurement ranges of the needed spectrometers. Right: Exemplary spectra of the applied AOTF for the visible wavelength range for control of up to 8 chosen lines simultaneously

Table B.2 Trap laser I: electrical and optical properties

Parameter	Value	Unit
Model	PL980P330J	–
Manufacturer	Thorlabs	–
Laser type	Diode laser	–
Max. output power	330	mW
Operating current	600	mA
Wave length	974–976	nm
Spectrum range	0.5–1	nm
Spectrum range (95% power)	<2	nm
Polarisation	Linear	–
Beam profile	Gaussian profile	–

Table B.3 Trap laser II: electrical and optical properties

Parameter	Value	Unit
Model	TKS-OTS series	–
Manufacturer	Manlight	–
Laser type	Fiber laser	–
Operating mode	CW-modulated	–
Max. output power	5	W
Range of the output power	10–100	%
Long-term stability	$<\pm 2$	%/1 std
Wave length	1060–1100	nm
Spectrum range	<0.25	nm
Polarisation	Linear	–
Polarisation extinction ratio	>18	db
Beam profile	Gaussian profile	–
Beam diameter collimator	2.2	mm
Beam quality M^2	<1.1	–

Table B.4 DHM (imaging) laser: electrical and optical properties

Parameter	Value	Unit
Model	LCM-S-111-10-L	–
Manufacturer	Laser-export	–
Laser type	Diode-pumped, frequency-doubled solid-state laser	–
Operating mode	CW-modulated	–
Max. output power	11	mW
Long-term stability	<2	%/8 std
Wave length	532	nm
Spectrum range	<0.00001	nm
Polarisation	Linear	–
Polarisation extinction ratio	>20	db
Beam profile	Gaussian profile	–
Beam diameter collimator	1.18	mm
Divergence	0.58	mrad
Beam quality M^2	1.05	–
Coherence length	>50	m

Table B.5 Wave front sensor: technical data

Parameter	Value	Unit
Model	WFS150-5C	–
Manufacturer	Thorlabs	–
Aperture	5.95×4.76	mm^2
Resolution	1280×1024	pixel
Pixel size	4.65×4.65	μm
Number of microlenses	39×31	–
Wave front-accuracy	$\lambda/15$ rms @ 633 nm	–
Wave front-sensitivity	$\lambda/15$ rms @ 633 nm	–
Dynamic range of the wave front	$>100\lambda$ @ 633 nm	–
Exposure time	79–65000	μs
Frame rate	15	Hz
Digitalisation	8	bit
Digital port	USB	–
Array of microlenses		
Substrate material	Quartz	
Wavelength range	300–1100	nm
Lens spacing	150	μm
Lens diameter	146	μm
Reflectivity	<25	%
Effective focal length	3.7	mm

B.1.2   Spatial Light Modulator (SLM)

The utilised spatial light modulator is one the recent developments of the company *Holoeye* (Berlin). The applied model *Pluto Phase Only Spatial Light Modulator (Reflective)* is controlled via a USB-interface. The PLUTO phase modulator models are based on reflective LCOS micro-displays are optimised to provide a phase shift range of typically 2π. The devices are packaged in an small housing to ensure an integration into optical setups and applications. The available PLUTO phase modulators comprise 2 versions, optimised for a broad wavelength band centred at 850 nm (NIR II) and a version specialised for the near infrared around 1064 nm (NIR).

Display Type:	Reflective LCOS (Phase Only)
Resolution:	1920 × 1080
Pixel Pitch:	8.0 μm
Fill Factor:	87%
Active Area	15.36 × 8.64 mm (0.7′ diagonal)
Addressing	8 Bit (256 Grey Levels)
Signal Formats	DVI - HDTV Resolution
Frame Rate	60 Hz

B.1.3 Spectrometre

For analyzing light spectra, compact computer controlled CCD spectrometers with a USB-interface are utilised, which are based on a folded Czerny-Turner setup.

A/D Konverter	14 bit
Detector	CCD-sensor, 2048 pixel
Wavelength range	200–1100 nm
spectral resolution	0,5 nm (depends on grating and aperture)
Integration time of CCD	3–65 ms

B.1.4 Light Power Metres

The measurement of the radiated light power P_L is performed with a power metre of *Coherent* (Type: *Fieldmaster*, measurement range (dependent on the chosen sensor): 400 nm to 1064 nm) which is based on a photo diode. The non linearity of the photo diode is compensated electronically by adjusting it manually by specification the wavelength. For measurements on LEDs, this non linearity is negligible. For the optical tweezer setup, another - but similar - analogue hand-held laser power metre console is utilzed (Thorlabs, PM101, sensor: S130C).

B.1.5 Technical Specifications of Applied Cameras

Monochrome CCD-Camera	**Q-imaging Retiga 2000RV**
Light-Sensitive Pixels	1.92 million; 1600×1200
Binning Modes	$2 \times 2, 4 \times 4, 8 \times 8$
ROI (Region of Interest)	single-pixel increment up to full res.
Exposure/Integration Control	$10\,\mu s$ to $17.9\,min$ in $1\,\mu s$ increments
Sensor Type	Kodak®KAI-2020 progressive-scan interline
Pixel Size	$7.4\,\mu m \times 7.4\,\mu m$
Linear Full Well	40,000e- (1×1 binning); 80,000e- (2×2 binning)
Read Noise	16e- (@20 MHz)
Dark Current	0.084e-/pix/s
Cooling Technology	3-stage Peltier cooling
Cooling Type	Down to $-30\,°C$, software controlled $1\,°C$ increments
Digital Output	12 bits
Readout Frequency	20, 10 & 5 MHz
Frame Rate	10fps full resolution @ 12 bits
Digital Interface	IEEE 1394 FireWire
External Trigger	TTL Input (optically coupled)
Trigger Types	Internal, Software, External
External Sync	TTL Output (optically coupled)
Gain Control	0.451 to 21.5 times
Offset Control	-2048 to 2047
Info website	http://www.qimaging.com/products/cameras/
Monochrome CCD-Camera	**Imaging Source DMK 41BU02**
Sens type	CCD
Frame rate	15, 7.5, 3.75 fps
Sensitivity	0.05 lx
Dynamic range	8 bit
SNR	ADC: 9 bit at $25\,°C$, gain 0 dB
Sensor specification	ICX205AL ICX205AL[271.91KB]
Type	progressive scan
Format	1/2'
Resolution	H:1360,V:1024, effective:1280×960
Pixel size	H: $4.65\,\mu m$, V: $4.65\,\mu m$
Shutter	1/10000 to 30 s
Gain	0 to 36 dB
Offset	0 to 511
Saturation	0 to 200 %
White balance	$-2\,dB$ to $+6\,dB$
Info website	http://www.theimagingsource.com/en_US/products

B.1.6   Piezo Actuator

For temporal phase shifting in the reference beam paths via the movement of a mirror in the different setups as well as for harmonisation of both interferometre beam paths, an innovative piezo actuator with long travel range and dual drive system from *Physik Instrumente* is applied (Table B.6).

B.1.7   Resolution Test Chart

Physical dimension: $50 \times 50 \times 1.5$ mm, chrome on glass (Table B.7 and Fig. B.2).

Table B.6 Technical specifications of utilised *PI* N-381.3A piezo actuators with NEXACT®linear drive. *Source* Information of manufacturer, www.pi.ws

Model Motion and positioning	N-381.3A, Drive type NEXACT®linear drive Mechanical properties
Travel range	30 mm
Stiffness in motion direction	2.4 N μm
Integrated sensor	Incremental linear encoder
Max. push/pull force (active)	10 N
Sensor resolution	20 nm*
Max. holding force (passive)	15 N
Travel range in analog mode	7 μm
Lateral force	10 N
Open-loop resolution	0.03 nm
Closed-loop resolution	20 nm
Max. velocity	10 mm/s

Table B.7 Line width of group elements of *Edmund* USAF-1951 test charts, specifications in μm, (*Source* Information of manufacturer)

Element	Group (Group 8, 9: *high resolution*)									
	0	1	2	3	4	5	6	7	8	9
1	500	250	125	62.5	31.3	15.6	7.8	3.9	2.0	0.9
2	446	223	111	55.7	27.9	13.9	7.0	3.5	1.7	0.8
3	397	198	99.2	49.5	24.8	12.4	6.2	3.1	1.5	0.7
4	355	176	88.3	44.2	22.1	11.0	5.5	2.8	1.4	—
5	314	157	78.7	39.4	20.0	9.8	4.9	2.5	1.2	—
6	281	140	70.1	35.0	17.5	8.8	4.4	2.2	1.1	—

Fig. B.2 USAF-1951
resolution test chart

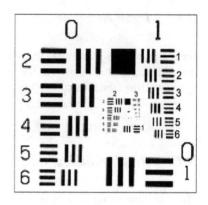

B.1.8   Ultra-Bright Light Emitting Diodes

Light emitting diodes (LEDs) exist in many designs and are used in various fields [2].
The main applications of LEDs are displays, lighting, remote controls, data transmission in optical fibers as well as motion sensors. One must distinguish between the following types of diodes : LED (Light-emitting diode), OLED (Organic Light-Emitting
Diode), LD (laser diode) and SLD (Super-Luminescence Diode). The structure and
principle of the LEDs are briefly described below.

LEDs are generally composed of a semiconductor crystal, which, for forming an
active region, is doped both positive and negative (in short: PN junction). First various
semiconductor layers are epitaxially applied on a wafer, which has the required
luminescence properties [2]. Frequently used metal and semiconductor components
especially in the case of ultra-bright LEDs are aluminumgalliumindiumphosphates
[2] (AlGaInP, Fig. B.4), galliumarsenphosphate (GaAsP) und indiumgalliumnitrite
(InGaN). The term ultra-bright LEDs designates no specific type, but only increased
efficiency and brightness achieved through new material compositions, which are
improved in the progressive development continuously. The doped, cube-shaped
semiconductor crystal is, in many LED types, positioned in a conical reflector pan.
In this way, a directional radiation is realised and luminous efficiency of the LED
increased (see Fig. B.3). The reflector can have a negative impact on the coherence
properties, which is of particular importance for the present study. The connection
to cathode and anode is done by a gold wire which is applied on the upper surface of
the semiconductor according to Fig. B.3. Subsequently, the diode is surrounded with
epoxy resin. The lens-shaped plastic body also favours a directional light emission.
Moreover, corrosion is prevented and the critical angle of total reflection at the chip
surface is reduced, thereby increasing the radiation emerging from the crystal.

The operating principle of an LED is based on a PN junction, which emits -
when applying current in the forward direction of polarity of the voltage and direct
transitions from the conduction band in the valence band of light (Fig. B.4). The
photons generated in this process result from recombination of electrons and holes.

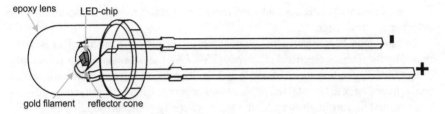

Fig. B.3 Layout of a typical 5 mm LED

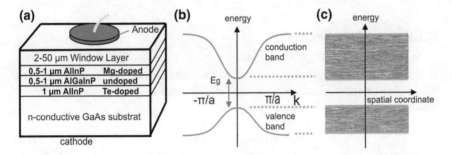

Fig. B.4 a Layer structure of a semiconductor chip of an ultra-bright LED, **b** Direct, light emitting transition within the band structure of a semiconductor. Band structure in k-space and as function of the spatial coordinate **c** of the crystal

The color of the emitted light depends on the size of the band gap as well as the material composition [3].

The emitted light output of an LED is proportional to the electric current flow I_e. The parameter I_e result to [4]

$$I_e(U, T) = q D_0 n_i^2 W e^{\frac{eU}{k_b T}} \quad . \tag{B.1}$$

where n_i is the intrinsic carrier density, W is the width of the space charge region, k_B is Boltzmann's constant, T the temperature, eV is the potential energy in the electric field, D_0 is a material-specific constant and q is the elementary charge.

The radiant power P_L of a light source is determined by the electric power P_{el} and the quantum efficiency ξ [4]:

$$P_L = P_{el} \cdot \xi = U \cdot I_e(U, T) \cdot \xi \quad . \tag{B.2}$$

The parameter ξ in LEDs is directly proportional to the fraction of recombining electron-hole pairs and depends on the parameters U, I and T. The quantum efficiency ξ is greatly dependent on temperature, since higher temperatures change the band structure and thus the energy transitions of the semiconductor [2]. Furthermore the quantum efficiency depends on material and design. It can also be enhanced by the use of intrinsic materials and such materials that are mostly transparent in the emitting spectral range.

Ultra-bright LEDs today achieve a quantum efficiency of up to 15%. This is determined by the internal electrical efficiency (75–87%), the internal quantum efficiency (about 90% depending on conditions) and the loss due to internal reflection, which can be traced back to the total reflection at the border crossing between semiconductor material and its surrounding material ($\approx$25% at epoxy resin surrounding).

For further details, please refer to the literature [3].

B.1.9   Lambert Emitters

To describe the radiation pattern of LEDs, the Lambert-radiator is often used as a model of an extended planar light source. This has a cosine-formed radiation intensity which depends on the viewing angle $\tilde{\alpha}$ [4]:

$$I_V(\tilde{\alpha}) = I_{V0} \cos(\tilde{\alpha}) \quad . \tag{B.3}$$

Here I_{V0} is the radiant intensity emitted perpendicular to a surface element. The beam intensity is defined as $I_V = \frac{d\tilde{\Phi}}{d\Omega}$ and denotes the radiation flux $\tilde{\Phi}$ attributable to the solid angle element $d\Omega$. A surface element dq of a light source emits a directional beam intensity $dI_V(\tilde{\alpha})$ (Fig. B.5). The corresponding radiance L, which is defined as

$$L(I_V(\tilde{\alpha})) = \frac{1}{\cos(\tilde{\alpha})} \frac{dI_V(\tilde{\alpha})}{dq} \quad , \tag{B.4}$$

is a measure of the density of energy dissipation. The factor $1/\cos(\tilde{\alpha})$ takes into account the size of the surface element decreasing with $\cos(\tilde{\alpha})$.

Substituting (B.3) into (B.4) shows that the Lambert emitters for each direction has the same radiation density $(L(I_V(\tilde{\alpha})) = \text{constant})$. This means that there are simple analytical relations for radiation density and radiation strength of the Lambert-Emitter.

Fig. B.5 Illustration of the cosine-formed radiation intensity depending on the viewing angle $I_V(\tilde{\alpha})$ in a solid angle element $d\Omega$

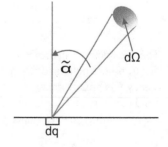

Fig. B.6 Classification of
the LED-spectra into the
standardised diagram for an
observer at 10° according to
CIE

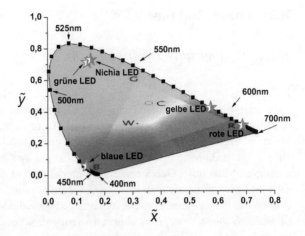

B.1.10   CIE-Classification of Light Sources

For a comparison of colours of different light sources a CIE-diagram is often used
since this type of diagram takes into account the specific colour impression of the
human eye. For a better visualisation this diagram consists of a projection of the three
dimensional $\tilde{X}\tilde{Y}\tilde{Z}$-colour space onto the $\tilde{X}\tilde{Y}$-plane. The $\tilde{X}$, $\tilde{Y}$ and $\tilde{Z}$-components are
determined with empirical sensitivity curves $\bar{x}_\lambda$, $\bar{y}_\lambda \bar{z}_\lambda$ (*colour matching functions*)
of the human eye ((B.5), [5]).

$$\tilde{X} = \Delta\lambda \sum_{\lambda=380\,nm}^{780\,nm} \bar{x}_\lambda \cdot B(\lambda); \quad \tilde{Y} = \Delta\lambda \sum_{\lambda=380\,nm}^{780\,nm} \bar{y}_\lambda \cdot B(\lambda); \quad \tilde{Z} = \Delta\lambda \sum_{\lambda=380\,nm}^{780\,nm} \bar{z}_\lambda \cdot B(\lambda) \quad \text{(B.5)}$$

This corresponds to the scalar product of the spectral function with the sensitivity
function of the eye, multiplied with the wave length interval $\Delta\lambda$ of the discrete given
values for $\bar{x}_\lambda$, $\bar{y}_\lambda \bar{z}_\lambda$[1] [5].

   Since the colour impression is evoked by the interaction of the XYZ-values, in a
CIE-diagram the relative values are plotted:

$$\tilde{x} = \frac{\tilde{X}}{\tilde{X}+\tilde{Y}+\tilde{Z}}; \quad \tilde{y} = \frac{\tilde{X}}{\tilde{X}+\tilde{Y}+\tilde{Z}}; \quad \tilde{z} = \frac{\tilde{X}}{\tilde{X}+\tilde{Y}+\tilde{Z}} \quad \text{(B.6)}$$

Therefore, the CIE diagram has the property that pure colours are located on the
boundary curve of the graph. In Fig. B.6 it is demonstrated that except the green
LED generate a pure colour impression.

---

[1]The value for $\Delta\lambda$ is typically 5 nm (CIE1938). CIE1971 and CIE1971 deliver values with 1 *nm*
interval. For further information it is referred to literature.

B.2   Theoretical Basics

B.2.1   Holography Throughout the Time

Holography is in fact known for a long time - since the first experiments by the Hungarian physicist Dennis Gabor in 1947 to improve the resolution power of an electron microscope. He has verified his basic theory using light waves instead of an electron beam and so the first hologram was created. It was an easily recognizable, simple figure showing many imperfections caused by the inappropriate light source available at that time. Gabor carried out model optical experiments to demonstrate the feasibility of the method. However, powerful sources of coherent light were not available at the time, and holography remained an optical paradox until the invention of lasers. Gabor's theory was almost 15 years ahead of it's time and the discovery he made was not really recognised until in the early sixties, when the laser as a light source with the required coherence was invented. Two scientists from University of Michigan (Emmett Leith and Juris Upatnieks) first utilised a laser to create a system that was able to reproduce a 3D image of an object. Thus, they manufactured the first diffuse-light hologram.

The notions "holography" and "hologram" were first applied by Dennis Gabor, who won the Nobel Prize for Physics in 1971. Holos is Greek for total or complete and graphein for to write. Holography denotes therefore the writing of the whole that describes the fact that not only the intensity of the light but also the direction of it's incidence is recorded using holography. Thus, a whole complex wave front resulting in a 3D image can be stored and read out. The word gramma has versatile meanings such as a unit. Therefore, a hologram represents the unity of the whole or a wholeness of the unity respectively, which means that the data storage is redundant since a small element is sufficient to reconstruct the recorded wave front, although the signal to noise ratio increases with the fraction of the original hologram.

Pieter J. van Heerden from Polaroid offered the first 3D holograms in 1963 in the form of data storage in three dimensions and the emergence of first holographic memories began. Since semiconductor chips rested cheaper throughout the time, holographic data storage is still rather an extraordinary solution when special requirements have to be fulfilled.

In microscopy, holography was deemed, a long time ago, to bring the capability of "seeing in 3D" microscopical objects. The capture of holograms was however so delicate that no real exploitation of the idea was pursued. In the field of biology and medicine, holography, applied to cell and tissue investigations in 3D, appeared as an attractive tool, but was rarely used in its early, non digital phase. The information nature of optics and holography is especially distinctly seen in digital holography. Wave field recorded in the form of a hologram in optical, radio frequency or acoustic holography, is represented in digital holography by a digital signal that carries the wave field information deprived of its physical casing.

Today the scientists' efforts are focused to commercialisation of holography where significant problems in miniaturisation and costs of the manufacturing process have to be faced. Therefore, many universities and various enterprises are still exploring better holographic materials and techniques.

The availability of digital cameras for recording intensity distributions and spatial light modulators highly supported the progress in holographic methods especially in the last two decades.

B.2.2   Maxwell Equations

The electromagnetic fields and the current and charge densities considered so far are real functions of space and time. When, as assumed in previous sections, time-harmonic fields are treated at a single frequency ω_0, one can define, for example, the electric field as

$$\xi(x, y, z, t) = \text{Re}[\mathbf{E}(x, y, z) \exp(j\omega_0 t)], \tag{B.7}$$

where $\mathbf{E}(x, y, z)$ is a vector field phasor corresponding to the time varying electric field $\xi(x, y, z, t)$. The phasor is complex in general as it has amplitude and phase information. The following phasors are present: $\mathbf{B}(x, y, z)$, $\mathbf{D}(x, y, z)$, $\mathbf{H}(x, y, z)$, $\mathbf{J}_c(x, y, z)$ and $\rho_v(x, y, z)$. With these phasors for the time-harmonic quantities and for a linear, isotropic, and homogeneous medium characterised by ε and μ, one can formulate Maxwell's equations as follows:

$$\nabla \cdot \mathbf{D} = \rho_v(x, y, z), \tag{B.8}$$

$$\nabla \cdot \mathbf{B} = 0, \tag{B.9}$$

$$\nabla \times \mathbf{E} = -j\omega_0 \mathbf{B}, \tag{B.10}$$

$$\nabla \times \mathbf{H} = \mathbf{J} = \mathbf{J}_c - j\omega_0 \mathbf{D}, \tag{B.11}$$

with $\mathbf{D} = \varepsilon \cdot \mathbf{E}$ and $\mathbf{E} = \mu \cdot \mathbf{H}$.

B.3   Wave Optics

In this Section, the *spatial frequency transfer function* for wave propagation is derived. Then the important *Fresnel diffraction* formula and the Fraunhofer Diffraction formula commonly used in Fourier optics is derived. it is started from the wave equation, expressed in Cartesian coordinates:

$$\frac{1}{v^2} \frac{\partial^2 \psi}{\partial t^2} = \frac{\partial^2 \psi}{\partial x^2} + \frac{\partial^2 \psi}{\partial y^2} + \frac{\partial^2 \psi}{\partial z^2} \tag{B.12}$$

It is now assumed that the wave function $\psi(x, y, z, t)$ comprises a *complex ampli-tude* $\psi_p(x, y, z)$ riding on a carrier of frequency ω_0 (ψ_p is a phasor in electrical engineering):

$$\psi(x, y, z, t) = \psi(x, y, z) \exp(j\omega_0 t). \tag{B.13}$$

Substituting (B.13) into (B.12), we get the *Helmholtz equation* for ψ_p,

$$\frac{\partial^2 \psi_p}{\partial x^2} + \frac{\partial^2 \psi_p}{\partial y^2} + \frac{\partial^2 \psi_p}{\partial z^2} + k_0^2 \psi_p = 0, \; k_0 = \frac{\omega_0}{v}. \tag{B.14}$$

At this point, we introduce 2-D *Fourier transform* in the next section and will there-after use Fourier transform to find the solution to (B.14) for a given initial condition.

B.3.1   Fourier Transform and Convolution

The two-dimensional (2-D) spatial Fourier transform of a square-integrable function $f(x, y)$ is given as

$$\begin{aligned} F(k_x, k_y) &= \int_{-\infty}^{\infty} \int_{-\infty}^{\infty} f(x, y) \exp(jk_x x + jk_y y) \mathrm{d}x \mathrm{d}y \\ &= \mathcal{F}_{xy}\{f(x, y)\} \end{aligned} \tag{B.15}$$

The inverse Fourier transform is

$$\begin{aligned} f(x, y) &= \frac{1}{4\pi^2} \int_{-\infty}^{\infty} \int_{-\infty}^{\infty} F(k_x, k_y) \exp(-jk_x x - jk_y y) \mathrm{d}k_x \mathrm{d}k_y \\ &= \mathcal{F}_{xy}^{-1}\{F(k_x, k_y)\} \end{aligned} \tag{B.16}$$

The definitions for the forward and backward transforms are consistent with the engi-neering convention for a travelling wave. In many optics applications, the function $f(x, y)$ represents the transverse profile of an electromagnetic or optical field at a plane z. Hence in (B.15) and (B.16), $f(x, y)$ and $F(k_x, k_y)$ have z as a parameter.

The usefulness of this transform lies in the fact that when substituted into the wave equation, one can reduce a three-dimensional partial differential equation (PDE) to a one-dimensional ordinary differential equation (ODE) for the spectral ampli-tude $F(k_x, k_y; z)$. The *convolution* $g(x, y)$ of two functions $g_1(x, y)$ and $g_2(x, y)$ is defined as

$$\begin{aligned} g(x, y) &= \int_{-\infty}^{\infty} \int_{-\infty}^{\infty} g_1(x', y') g_2(x - x', y - y') \mathrm{d}x' \mathrm{d}y' \\ &= g_1(x, y) * g_2(x, y). \end{aligned} \tag{B.17}$$

It can be readily shown that the Fourier transform $G(k_x, k_y)$ of $g(x, y)$ is related to the Fourier transforms $G_{1,2}(k_x, k_y)$ of $g_{1,2}(x, y)$ as

$$G(k_x, k_y) = G_1(k_x, k_y)G_2(k_x, k_y). \tag{B.18}$$

B.3.2 Spatial Frequency Transfer Function and Spatial Impulse Response of Propagation

By taking the 2-D Fourier transform, i.e., $\mathcal{F}_{xy}$, of (B.14) and upon some manipulations, we have

$$\frac{d^2\Psi_p}{dz^2} + k_0^2 \left(1 - \frac{k_x^2}{k_0^2} - \frac{k_y^2}{k_0^2}\right)\Psi_p = 0,$$

where $\Psi_p(k_x, k_y; z)$ is the Fourier transform of $\psi_p(x, y, z)$. We now readily solve the above equation to get

$$\Psi_p(k_x, k_y; z) = \Psi_{p0}(k_x, k_y)\exp[-jk_0\sqrt{1 - k_x^2/k_0^2 - k_y/k_0^2}z], \tag{B.19}$$

where $\Psi_{p0}(k_x, k_y) = \Psi_p(k_x, k_y; z = 0)$

$$= \mathcal{F}_{xy}\{\psi_p(x, y, z = 0)\} = \mathcal{F}_{xy}\{\psi_{p0}(x, y)\}.$$

We can interpret (B.19) in the following way: Consider a linear system with $\psi_{p0}(k_x, k_y)$ as its input spectrum (i.e., at $z = 0$) and where the output spectrum is $\psi_p(k_x, k_y; z)$. Then, the spatial frequency response of the system is give by

$$\frac{\Psi_p(k_x, k_y; z)}{\Psi_{p0}(k_x, k_y)} = \mathcal{H}(k_x, k_y; z)$$

$$= [-jk_0\sqrt{1 - k_x^2/k_0^2 - k_y^2/k_0^2}z]. \tag{B.20}$$

We will call $\mathcal{H}(k_x, k_y; z)$ the *spatial frequency transfer function* of propagation of light through a distance z in the medium. To find the field distribution at z in the spatial domain, we take the inverse Fourier transform of (B.19):

$$\psi_p(x, y, z) = \mathcal{F}_{xy}^{-1}\{\Psi_p(k_x, k_y; z)\}$$

$$= \frac{1}{4\pi^2}\int\int \Psi_{p0}(k_x, k_y)\exp[-jk_0\sqrt{1 - k_x^2/k_0^2 - k_y^2/k_0^2}z] \tag{B.21}$$

$$\times \exp[-k_x x - jk_y y]dk_x dk_y.$$

Now, by substituting $\Psi_{p0}(k_x, k_y) = \mathcal{F}_{xy}\{\psi_{p0}(x, y)\}$ into (B.21), we can express $\psi_p(x, y, z)$ as

$$\psi_p(x, y, z) = \int \int \psi_{p0}(x', y')G(x - x', y - y'; z)dx'dy'$$
$$= \psi_{p0}(x, y) * G(x, y; z).$$
(B.22)

where

$$G(x, y; z) = \frac{1}{4\pi^2} \int \int \exp[-jk_0\sqrt{1 - k_x^2/k_0^2 - k_y^2/k_0^2}z]$$
$$\times \exp[-jk_x x - jk_y y]dk_x dk_y$$

The result of (B.22) indicates that $G(x, y; z)$ is the *spatial impulse response of propagation* of the system. By changing of variables: $x = r\cos\theta$, $y = r\sin\theta$, $k_x = \rho\cos\phi$, and $k_y = \rho\sin\phi$, $G(x, y; z)$ can be evaluated as

$$G(x, y; z) = \frac{jk_0 \exp(-jk_0\sqrt{r^2 + z^2})}{2\pi\sqrt{r^2 + z^2}}$$
$$\times \frac{z}{\sqrt{r^2 + z^2}}\left(1 + \frac{1}{jk_0\sqrt{r^2 + z^2}}\right).$$
(B.23)

One may observe the following:

(1) For $z \gg \lambda_0 = 2\pi/k_0$, i.e., we observe the field distribution many wavelengths away from the diffracting aperture, we have $(1 + \frac{1}{jk_0\sqrt{r^2z^2}}) \approx 1$.

(2) $\frac{z}{\sqrt{r^2+z^2}} = \cos\Phi$, where $\cos\Phi$ is called the *obliquity factor* and Φ is the angle between the positive z-axis and the line passing through the origin of the coordinates. Now, using the binomial expansion, the factor $\sqrt{r^2 + z^2} = \sqrt{x^2 + y^2 + z^2} \approx z + \frac{x^2+y^2}{2z}$, provided $x^2 + y^2 \ll z^2$. This condition is called the *paraxial approximation*, which leads to $\cos\Phi \approx 1$. If the condition is used in the more sensitive phase term and only used the first expansion term in the less sensitive denominators of the first and second terms of (B.23), $G(x, y; z)$ becomes the so-called *free-space spatial impulse response*, $h(x, y; z)$, in Fourier Optics:

$$h(x, y; z) = \exp(-jk_z z)\frac{jk_0}{2\pi z}\exp\left[\frac{-jk_0(x^2 + y^2)}{2z}\right].$$
(B.24)

By taking the 2-D Fourier transform of $h(x, y; z)$, we have

$$H(k_x, k_y; z) = \mathcal{F}_{xy}\{h(x, y; z)\}$$
$$= \exp(-jk_0 z)\exp\left[\frac{j(k_x^2 + k_y^2)z}{2k_0}\right].$$
(B.25)

$$\psi_{p0}(x, y) \longrightarrow \boxed{h(x, y; z)} \longrightarrow \psi_p(x, y, z)$$

Fig. B.7 Block diagram of wave propagation in Fourier optics

$H(k_x, k_y; z)$ is called the *spatial frequency response* in Fourier Optics. Indeed, we can derive (B.25) directly if we assume that $k_x^2 + k_y^2 \ll k_0^2$, meaning that the x and y components of the propagation vector of a wave are relatively small, we have, from (B.20)

$$\begin{aligned}
\frac{\Psi_p(k_x, k_y; z)}{\Psi_{p0}(k_x, k_y)} &= \mathcal{H}(k_x, k_y; z) \\
&= \exp\left[-jk_0\sqrt{1 - k_x^2/k_0^2 - k_y^2/k_0^2}\,z\right] \\
&\simeq \exp(-jk_0 z)\exp\left[\frac{j(k_x^2 + k_y^2)z}{2k_0}\right] \\
&= H(k_x, k_y; z).
\end{aligned} \tag{B.26}$$

If (B.24) is now used in (B.22), we obtain

$$\begin{aligned}
\psi_p(x, y, z) &= \psi_{p0}(x, y) * h(x, y; z) \\
&= \exp(-jk_0 z)\frac{jk_0}{2\pi z} \int\int \psi_{p0}(x', y') \\
&\quad \times \exp\left[\frac{-jk_0}{2z}\left((x - x')^2 + (y - y')^2\right)\right] dx'dy'.
\end{aligned} \tag{B.27}$$

Equation (B.27) is called the *Fresnel diffraction formula* and describes the Fresnel diffraction of a beam during propagation and having an arbitrary initial complex profile $\psi_{p0}(x, y)$. The input and output planes have primed and unprimed coordinate systems, respectively. Figure B.7 shows a block-diagram relating the input and output planes. To obtain the output field distribution $\psi_p(x, y, z)$ at a distance z away from the input (the location of the diffracting screen), we need to convolve the input field distribution $\psi_p(x, y, z)$ with the spatial impulse response $h(x, y; z)$.

B.3.3 Examples of Fresnel Diffraction

Example 1: Point source

A point source is represented by $\psi_{p0} = \delta(x)\delta(y)$. By (B.27), the complex field at a distance z away is given by

$$\psi_p(x, y, z) = [\delta(x)\delta(y)] * h(x, y, z)$$

$$= \frac{jk_0}{2\pi z} \exp\left[-jk_0 z - \frac{jk_0(x^2 + y^2)}{2z}\right]. \tag{B.28}$$

This expression is the paraxial approximation to a *diverging spherical wave*. Now, by considering the argument of the exponent in (B.28), we see that using the binomial expansion previously used, we can write

$$psi_p(x, y, z) \simeq \frac{jk_0}{2\pi z} \exp\left(-jk_0[x^2 + y^2 + z^2]^{1/2}\right)$$

$$\simeq \frac{jk_0}{2\pi z} \exp(-jk_0 R), \tag{B.29}$$

which corresponds to a diverging spherical wave.

Example 2: Plane Wave

For a plane wave, we write $\psi_{p0}(x, y) = 1$. Then $\Psi_{p0}(k_x, k_y) = 4\pi^2\delta(k_x)\delta(k_y)$. Using (B.25), we have

$$\Psi_p(k_x, k_y, z) = 4\pi^2\delta(k_x)\delta(k_y)\exp(-jk_0 z)\exp\left[\frac{j(k_x^2 + k_y^2)z}{2k_0}\right] \tag{B.30}$$

$$= 4\pi^2\delta(k_x)\delta(k_y)\exp(-jk_0 z)$$

Hence,

$$\psi_p(x, y, z) = \exp(-jk_0 z)$$

As the plane wave travels, it only acquires phase shift and is undiffracted, as expected.

B.3.4 Fraunhofer Diffraction

When we examine the Fresnel diffraction pattern, which is calculated through the Fresnel diffraction formula (B.27), the range of applicability of this formula is from distances not too close to the source, typically from about 10 times the wavelength in practice. In this section we examine a method of calculating the diffraction pattern at distances far away from the source or aperture. More precisely, we observe in the *far field*, that is,

$$\frac{k_0(x'^2 + y'^2)_{max}}{2} = z_R \ll z, \tag{B.31}$$

where z_R is the *Rayleigh range*, then the value of the exponent $\exp[-jk_0(x'^2 + y'^2)]_{max}/2z$ is approximately unity over the input plane (x', y'). Under this assumption, which is called the *Fraunhofer approximation*, (B.27) becomes

$$\psi_p(x, y, z) = \exp(-jk_0 z)\frac{jk_0}{2\pi z} \exp\left[\frac{-jk_0}{2z}(x^2 + y^2)\right]$$

$$\times \int\int \psi_{p0}(x', y') \exp\left[\frac{-jk_0}{z}(xx' + yy')\right] dx' dy'.$$

$$= \exp(-jk_0 z)\frac{jk_0}{2\pi z} \exp\left[\frac{-jk_0}{2z}(x^2 + y^2)\right]$$

$$\times \mathcal{F}_{xy}\{\psi_{p0}(x, y)\}\Big|_{\substack{k_x=k_0 x/z \\ k_y=k_0 y/z}}$$

$$\text{(B.32)}$$

Equation (B.32) is the *Fraunhofer diffraction formula* and is the limiting case of the Fresnel diffraction studied earlier. The first exponential in (B.32) is the result of the phase change due to propagation, whereas the second exponential indicates a phase curvature that is quadratic in nature. Note that if we are treating diffraction of red light ($\lambda = 0.6328\,\mu m$) and the maximum dimensions on the input plane are 1 mm, the distance to observe the far field is then, according to (B.31), $z \gg 5\,m$.

Example: Fraunhofer Diffraction of a Slit of Finite Width

The complex amplitude of a slit illuminated by a plane wave of unity amplitude is represented by $\psi_{p0}(x, y) = \text{rect}\left(\frac{x}{l_x}\right)$ at the exit of the slit of width l_x along the x-direction. Note that because we are usually interested in diffracted intensities (i.e., $|\psi_p|^2$), the exponentials in (B.32) drop out. Furthermore, the other term besides the Fourier transform, namely, $(k_0/2\pi z)$, simply acts as a weighting factor. The intensity profile depends on the Fourier transform, and we will therefore concentrate only on this unless otherwise stated. Using

$$\mathcal{F}_{xy}\left\{\text{rect}\left(\frac{x}{l_x}\right)\right\} = l_x \cdot \text{sinc}\left(\frac{l_x k_x}{2\pi}\right)\sqrt{2\pi}\delta(k_y),$$

and from (B.32), we have

$$\psi_p(x, y; z) \propto l_x \cdot \text{sinc}\left(\frac{l_x k_0 x}{2\pi z}\right) \cdot \delta\left(\frac{k_0 y}{z}\right) \tag{B.33}$$

Since there is no variation along y, we plot the normalised intensity $I(x)/I(0) = \text{sinc}^2\left(\frac{l_x k_0 x}{2\pi z}\right)$ along x only as shown in Fig. B.8.

We observe that the first zero of the *sinc*-function occurs at $x = \pm 2\pi z/l_x k_0 = \pm \lambda_0 z/l_x$, and that the diffraction angle $\theta_{\text{spread}} \simeq \lambda_0/l_x$ during diffraction.

In fact, we can simply find the spread angle from a quantum mechanical point of view. Consider light emanating from an aperture of width l_x, as shown in Fig. B.9.

Quantum mechanics relates the minimum uncertainty in position Δx of a quantum to the uncertainty in its momentum Δp_x according to

$$\Delta x \Delta p_x \sim h. \tag{B.34}$$

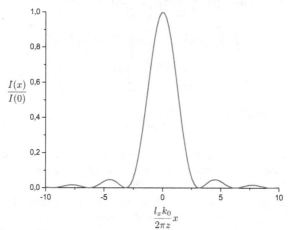

Fig. B.8 Fraunhofer diffraction pattern of a slit

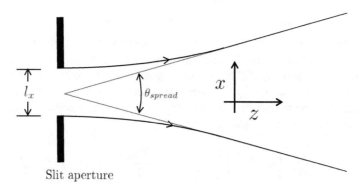

Fig. B.9 Geometry for determination of the diffraction angle θ_{spread} during diffraction

Now, in our problem $\Delta x = l_x$, because the *quantum* of light can emerge from any point on the aperture. Hence, by (B.34),

$$\Delta p_x \sim \frac{h}{l_x}.$$

We define the angle of spread θ_{spread}, assumed small, as

$$\theta_{\text{spread}} \sim \frac{\Delta p_x}{p_z} \sim \frac{\Delta p_x}{p_0}$$

where p, and p_0 represent z component of the momentum, and the momentum of the quantum, respectively. But $p_0 = \hbar k_0$, where k_0 is the propagation constant; hence,

$$\theta_{\text{spread}} \sim \frac{2\pi}{k_0 l_x} = \frac{\lambda_0}{l_x},$$

where λ_0 denotes the wavelength in the medium of propagation of the light. Thus, the angle of spread is inversely proportional to the aperture width as predicted previously using Fraunhofer diffraction.

B.3.5 Fourier Transforming Property of Ideal Lenses

Since a lens is a phase object, and for an ideal focusing lens of focal length f, its phase transformation function, $t_f(x, y)$, is given by

$$t_f(x, y) = \exp\left[j\frac{k_0}{2f}(x^2 + y^2)\right]. \tag{B.35}$$

The reason for this is that for a uniform plane wave incident upon the lens, the wave front behind the lens is a converging spherical wave (for $f > 0$) that converges ideally to a point source a distance $z = f$ behind the lens. Upon comparing with the paraxial approximation for a diverging spherical wave, as given by (B.28), (B.35) readily follows for an ideal thin lens (thickness of the lens being zero).

In the following the effect of placing a transparency $t(x, y)$ against the ideal lens is investigated, as shown in Fig. B.10. In general, $t(x, y)$ is a complex function such that if a complex field $\psi_p(x, y)$ is incident on it, the field immediately behind the transparency-lens combination is

Fig. B.10 Transparency in front of an ideal lens under complex field illumination

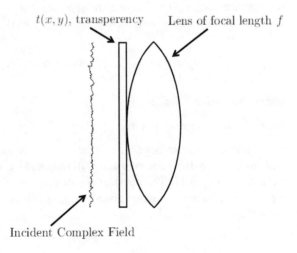

$t(x, y)$, transperency   Lens of focal length f

Incident Complex Field

$$\psi_p(x, y)t(x, y)t_f(x, y) = \psi_p(x, y)t(x, y) \exp\left[j\frac{k_0}{2f}(x^2 + y^2)\right],$$

where we have assumed that the transparency is infinitely thin, as is the case for the ideal lens. Then, for brevity, under illuminated by a unit amplitude plane wave, i.e, $\psi_p(x, y) = 1$, the field immediately behind the combination is given by $t(x, y) \exp\left[j\frac{k_0}{2f}(x^2 + y^2)\right]$. We then find the field distribution at a distance $z = f$ by using the Fresnel diffraction formula, (B.27), as

$$
\begin{aligned}
\psi_p(x, y, z = f) &= \exp(-jk_0 f)\frac{jk_0}{2\pi f} \exp\left[\frac{-jk_0}{2f}(x^2 + y^2)\right] \\
&\quad \times \int\int t(x', y') \exp\left[j\frac{k_0}{f}(xx' + yy')\right] dx'dy' \\
&= \exp(-jk_0 f)\frac{jk_0}{2\pi f} \exp\left[\frac{-jk_0}{2f}(x^2 + y^2)\right] \\
&\quad \times \mathcal{F}_{xy}\{t(x, y)\}\Big|_{\substack{k_x=k_0x/f \\ k_y=k_0y/f}}
\end{aligned}
\tag{B.36}
$$

where x and y denote the transverse coordinates at $z = f$. Hence, the complex field on the focal plane ($z = f$) is proportional to the Fourier transform of $t(x, y)$ but with a phase curvature. Note that if $t(x, y) = 1$, i.e., the transparency is totally clear, we have $\psi_p(x, y, z = f) \propto \delta(x, y)$, which corresponds to the focusing of a plane wave by a lens. Note that, for an ideal divergent lens, its phase transformation function is given by $\exp[-j\frac{k_0}{2f}(x^2 + y^2)]$.

All physical lenses have finite apertures and we can model this physical situation as a lens with an infinite aperture followed immediately by a transparency described by what is called the *pupil function* $p_f(x, y)$ of the lens. Typical pupil functions are rect$(x/X, y/Y)$ or circ(r/r_0), where X, Y, and r_0 are some constants, and $r = (x^2 + y^2)^{\frac{1}{2}}$ and that circ(r/r_0) denotes a value 1 within a circle of radius r_0 and 0 otherwise. Hence, if we have a transparency $t(x, y)$ against a lens with a finite aperture, the field at the back focal plane of the lens is given by

$$\psi_p(x, y, z = f) \propto \mathcal{F}_{xy}\{t(x, y)p_f(x, y)\}\Big|_{\substack{k_x=k_0x/f \\ k_y=k_0y/f}} \tag{B.37}$$

under plane wave illumination.

Example: Transparency in Front of a Lens

Suppose that a transparency $t(x, y)$ is located at a distance do in front of a convex lens with an infinitely large aperture and is illuminated by a plane wave of unit strength as shown in Fig. B.11. The physical situation is shown in Fig. B.11a, which can be represented by a block diagram given by Fig. B.11b. According to the block diagram, we write

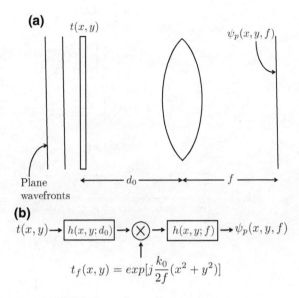

Fig. B.11 Plane-wave illumination of a transparency $t(x, y)$ located a distance d in front of a converging lens of focal length f: **a** Physical situation **b** Block diagram

$$\psi_p(x, y, f) = \{[t(x, y) * h(x, y; d_0)]t_f(x, y)\} * h(x, y; f), \tag{B.38}$$

which can be evaluated to become

$$\psi_p(x, y, f) = \frac{jk_0}{2\pi f} \exp(-jk_0(d_0 + f)) \exp\left[-\frac{jk_0}{2f}(1 - \frac{d_0}{f})(x^2 + y^2)\right]$$

$$\times \mathcal{F}_{xy}\{t(x, y)\}\bigg|_{\substack{k_x = k_0 x/f \\ k_y = k_0 y/f}} \tag{B.39}$$

Note that, as in (B.36), a phase curvature factor again precedes the Fourier transform, but vanishes for the special case $d_0 = f$. Thus, when the transparency is placed in the front focal plane of the convex lens the phase curvature disappears, and we recover the exact Fourier transform on the back focal plane. Fourier processing on an "input" transparency located on the front focal plane may now be performed on the back focal plane. Thus, a lens brings the Fraunhofer diffraction pattern (usually seen in the far field) to its back focal plane through the quadratic phase transformation. This is the essence of Fourier optics to perform coherent image processing.

B.3.6 Synopsis of Further Holographic Reconstruction Algorithms

In the following alternative approaches are summarised for holographic reconstruction according to [6]. The nomenclature of the variables is the same than in Chap. 2.

$$z \cdot \mathcal{F}^{-1}\{h \cdot r \cdot w\} \tag{B.40}$$

$$w(k, l) = \exp\left\{\frac{i\pi}{d'\lambda}(k^2\Delta\xi^2 + l^2\Delta\eta^2)\right\} \tag{B.41}$$

$$z(n, m) = \exp\left\{\frac{-i\pi d'\lambda}{N^2}\left(\frac{n^2}{\Delta\xi^2} + \frac{m^2}{\Delta\eta^2}\right)\right\} \tag{B.42}$$

$$\mathcal{F}^{-1}\{\mathcal{F}\{h \cdot r\} \cdot \mathcal{F}\{g\}\} \tag{B.43}$$

$$g(k, l) = \frac{1}{i\lambda}\frac{\exp\left\{\frac{2\pi i}{\lambda}\sqrt{d'^2 + (k - N/2)^2\Delta\xi^2 + (l - N/2)^2\Delta\eta^2}\right\}}{\sqrt{d'^2 + (k - N/2)^2\Delta\xi^2 + (l - N/2)^2\Delta\eta^2}} \tag{B.44}$$

$$\mathcal{F}^{-1}\{\mathcal{F}\{h \cdot r\} \cdot G\} \tag{B.45}$$

$$G(n, m) = \exp\left\{\frac{2\pi i d'}{\lambda}\sqrt{1 - \frac{\lambda^2\left(n + \frac{N^2\Delta\xi^2}{2d'\lambda}\right)^2}{N^2\Delta\xi^2} - \frac{\lambda^2\left(m + \frac{N^2\Delta\eta^2}{2d'\lambda}\right)^2}{N^2\Delta\eta^2}}\right\} \tag{B.46}$$

$$\mathcal{F}^{-1}\{\mathcal{F}\{h \cdot r\} \cdot \mathcal{F}\{g_F\}\} \tag{B.47}$$

$$g_F(k, l) = \frac{\exp(id'2\pi/\lambda)}{i\lambda d'}\exp\left\{\frac{i\pi}{\lambda d'}\left[(k - N/2)^2\Delta\xi^2 + (l - N/2)^2\Delta\eta^2\right]\right\} \tag{B.48}$$

$$\mathcal{F}^{-1}\{\mathcal{F}\{h \cdot r\} \cdot G_F\} \tag{B.49}$$

$$G_F(n, m) = \exp\left\{i\pi d'\left[\frac{2}{\lambda} - \lambda\left(\frac{n}{N\Delta\xi} + \frac{N\Delta\xi}{2d'\lambda}\right)^2 - \lambda\left(\frac{m}{N\Delta\eta} + \frac{N\Delta\eta}{2d'\lambda}\right)^2\right]\right\} \tag{B.50}$$

B.3.7   Mathematical Approximation of the Refractive Index

For a consideration of dispersion effects, a precise knowledge of the refractive index in the dispersive medium is required. In transparent media, the absorption coefficient $\tilde{\kappa}$ for the visible spectrum is very small compared to the real part n' and can therefore be neglected ($n \approx n'$). In general, the refractive index $n(\lambda)$ of most glasses is given in literature at a wavelength of 587.56 nm, which is sometimes indicated by n_d.

For the representation of the refractive index as a function which can be used for analytical calculations, a variety of approximations exist, which are mostly based on the theory of classical electrodynamics [7].[2] The Sellmeyer formula (B.51) is particularly useful for the theoretical description of the dispersion and the effects resulting from it since a numerical derivative of the refractive index leads to noise at higher orders.

$$n(\lambda) = \sqrt{1 + \sum_{i=1} \frac{B_i \lambda^2}{\lambda^2 - C_i}}, \quad (i \in \{1, 2, 3 \ldots\}) \quad, \tag{B.51}$$

with C_i and B_i as adjustable parameters. The absorption resonances are given by $\sqrt{C_i}$. At those positions there are poles of the Sellmeyer equation, which is the reason why it yields unphysical values of $n = \pm\infty$. To determine the Sellmeyer coefficients, the refractive indices of three standard wavelengths are required. These are usually the red cadmium line at 644 nm, the mercury line at 546 nm, and the blue cadmium line at 480 nm. In the proximity of the absorption resonances, a more accurate modelling of the refractive index should be used such as those of Helmholtz. There are a variety of modified Sellmeyer equations whose extensions expand the scope of the equation, e.g. for allowing to take into account the temperature dependence of the refractive index. At wavelengths far from the absorption maxima the refractive index can be further approximated by $n = \sqrt{1 + \sum_i B_i} \approx \sqrt{\epsilon_r}$, where ϵ_r denotes the relative dielectric constant. All in all an accuracy of $< 1 \cdot 10^{-5}$ is achievable in the visible spectral range by an approximation with the Sellmeyer formula. The Sellmeyer coefficients B_i and c_i, as provided by glass manufacturers, refer to normal room temperature. The treatment of the temperature dependence is discussed in detail in [7] which has shown to be negligible for the experiments described here.

B.3.8 Zernike Polynoms

Orthogonal functions arise naturally as eigenfunctions of Green's functions of differential equations and Fredholm integral equations. In optics, Zernike polynomials [8, 9] are mainly used in the analysis of interferogram fringes and for minimizing aberrations when manufacturing optical elements. In general, they are suitable for describing wavefronts within an circular aperture. Among their many uses, they give also a generalised Fourier expansion of a function [10]. Since the Zernike polynomials have been extensively discussed in literature, a complete presentation of their characteristics is omitted. Then, in what follows, emphasis is put on a few specific properties relevant to this work.

Several slightly different Zernike polynomial normalisations have been used in literature. Without loss in generality, here only the original normalisation (and corresponding definition) of the Zernike polynomials is given. The Zernike functions in Fig. B.12 are a product of the Zernike radial polynomials in combination with *sine*- and *cosine*-functions.

---

[2]The formula was derived by Sellmeyer in 1871. His work arose from the work of Augustin Cauchy on Cauchy's equation for modeling the dispersion.

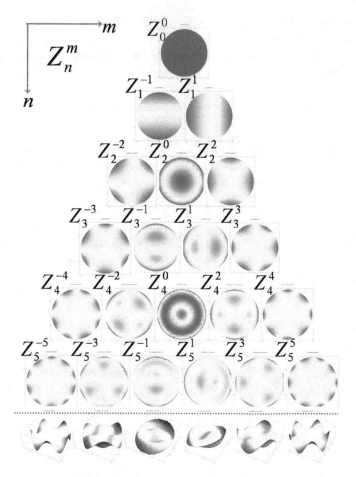

Fig. B.12 Zernike circle polynomials pyramid showing a plot for each polynomial. The standard deviation of each polynomial is one wave

$$\left\{ \begin{array}{c} Z_n^m(r, \theta) \\ Z_n^{-m}(r, \theta) \end{array} \right\} = R_n^m(r) \left\{ \begin{array}{c} \sin m\theta \\ \cos m\theta \end{array} \right\} \tag{B.52}$$

The index $n = 0, 1, 2, \ldots$ is named the degree of the function or polynomial, while $m = -n$ to $+n$, with $(n - m)$ even, is called the order. The radial polynomials are normally defined utilizing their series representation as a finite sum of powers of r^2:

$$R_n^m(r) = \sum_{k=0}^{(n-m)/2} \frac{(-1)^k (n - k)!}{k! \left(\frac{n+m}{2} - k \right)! \left(\frac{n-m}{2} - k \right)!} r^{n-2k} \quad n = 0, 1, 2, \ldots (n - m) = \pm 2k; \; k \in \mathbb{N}^0 \tag{B.53}$$

The low mode number functions are used so frequently for analysing optical data that many are given common names (see Table B.8). These functions are useful as a

Table B.8 Orthonormal Zernike circle polynomials $Z_j(\theta\rho)$ with the polynomial number j, the radial degree n and the azimuthal frequency m. In this order, an even j corresponds to a symmetric polynomial varying as $\cos(m\theta)$ and an odd j to an asymmetric polynomial varying as $\sin(m\theta)$

j	n	m	$Z_j(\theta\rho)$	Aberration name
1	0	0	1	Piston
2	1	1	$2\rho\cos\theta$	x tilt
3	1	1	$2\rho\sin\theta$	y tilt
4	2	0	$\sqrt{3}(2\rho^2 - 1)$	Defocus
5	2	2	$\sqrt{6}\rho^2\sin 2\theta$	Primary astigmatism at 45°
6	2	2	$\sqrt{6}\rho^2\cos 2\theta$	Primary astigmatism at 0°
7	3	1	$\sqrt{8}(3\rho^3 - 2\rho)\sin\theta$	Primary y coma
8	3	1	$\sqrt{8}(3\rho^3 - 2\rho)\cos\theta$	Primary x coma
9	3	3	$\sqrt{8}(3\rho^3)\sin 3\theta$	
10	3	3	$\sqrt{8}(3\rho^3)\cos 3\theta$	
11	4	0	$\sqrt{5}(6\rho^4 - 6\rho^2 + 1)$	Primary spherical
12	4	2	$\sqrt{10}(4\rho^4 - 3d\rho^2)\cos 2\theta$	Secondary astigmatism at 0°
13	4	2	$\sqrt{10}(4\rho^4 - 3d\rho^2)\sin 2\theta$	Secondary astigmatism at 0°
14	4	4	$\sqrt{10}\rho^4\cos 4\theta$	
15	4	4	$\sqrt{10}\rho^4\sin 4\theta$	
16	5	1	$\sqrt{12}(10\rho^5 - 12\rho^3 + 3\rho)\cos\theta$	Secondary x coma
17	5	1	$\sqrt{12}(10\rho^5 - 12\rho^3 + 3\rho)\sin\theta$	Secondary y coma
18	5	3	$\sqrt{12}(5\rho^5 - 4\rho^3)\cos 3\theta$	
19	5	3	$\sqrt{12}(5\rho^5 - 4\rho^3)\sin 3\theta$	
20	5	5	$\sqrt{12}\rho^5\cos 5\theta$	
21	5	5	$\sqrt{12}\rho^5\sin 5\theta$	
22	6	0	$\sqrt{7}(20\rho^6 - 30\rho^4 + 12\rho^2 - 1)$	Secondary spherical
23	6	2	$\sqrt{14}(15\rho^6 - 20\rho^4 + 6\rho^2)\sin 2\theta$	Tertiary astigmatism at 45°
24	6	2	$\sqrt{14}(15\rho^6 - 20\rho^4 + 6\rho^2)\cos 2\theta$	Tertiary astigmatism at 0°
25	6	4	$\sqrt{14}(6\rho^6 - 5\rho^4)\sin 4\theta$	
26	6	4	$\sqrt{14}(6\rho^6 - 5\rho^4)\cos 4\theta$	
27	6	6	$\sqrt{14}\rho^6\sin 6\theta$	
28	6	6	$\sqrt{14}\rho^6\cos 6\theta$	

basis for decomposing complex functions or data because they are orthogonal over the unit circle, meaning that they satisfy the following relation:

$$\int_0^1 \int_0^{2\pi} Z_n^m(r, \theta)Z_{n'}^{m'}(r, \theta)r\mathrm{d}r\mathrm{d}\theta = \frac{1}{2n + 1}\delta_{nn'}\delta_{mm'} \tag{B.54}$$

The factor $(2n + 1) - 1$ on the right is commonly used as a normalisation constant for the functions. Applying this orthogonality, any function $f(r, \theta)$ defined on the circle can be transformed to a sum of Zernike modes, analogous as *sine* and *cosine* functions are used in familiar 1-D Fourier transforms.

$$f(r, \theta) = \sum_{n=0}^{\infty} \sum_{m=-n}^{n} a_{nm} Z_n^m(r, \theta) \tag{B.55}$$

By representing data in this way one can compress a sophisticated structural deformation or aberration in terms of a small number of coefficients associated with the dominant Zernike modes. The coefficients of the last equation are evaluated by inverting the relation,

$$a_{nm} = \int_0^1 \int_0^{2\pi} f(r, \theta) Z_{n'}^{m'}(r, \theta) r \, dr \, d\theta : \tag{B.56}$$

The a_{nm} can be calculated directly from this equation if $f(r, \theta)$ is a known function, or computed numerically if f is a set of a measured surface. In the latter case, the formal method to compute these integrals is to use a numerical quadrature routine.

B.3.9   Forbes Polynoms

Alternatively to Zernike polynomials described in Sect. B.3.8, the Forbes Polynomials represent a tool for the characterisation of optical surface shapes or wavefronts [11], which is useful for various applications including optical fabrication, testing, and patenting, as well as design. The advantage of this representation is the more intuitive handling of the coefficients in case a user has to adapt these dynamically. Usually rotationally symmetric surface shapes are mathematically described in terms of their deviation from the sagittal representation (or known as "sag") of a close-fitting conic:

$$z(\rho) = \frac{c\rho^2}{\left(1 + \sqrt{1 - \epsilon c^2 \rho^2}\right)} + \sum_{m=0}^{M} a_m \rho^{2m+4} \tag{B.57}$$

where $\{z, r, \varphi\}$ are standard cylindrical polar coordinates. Here, the parameter c is the paraxial curvature of the surface and ϵ the conic parameter. Only a handful of terms are typically retained in the added polynomial, but it is increasingly common to see this number grow in order to characterize a desired shape with sufficient accuracy. For the purposes of fabrication and testing, this characterisation of the surface shape is completed by specification of the aperture size, i.e. a value for r_{max} where (B.57) is then valid over $0 < r < r_{max}$.

The form described bz (B.57) however often leads to numerical problems and therefore a better form for the added polynomial terms, that describe the deviation from an ideal conic has to be found. This is where the Forbes Polynomials come into play. They are chosen as new basis for the second term. Equation (B.57) is rewritten as

$$z(\rho) = \frac{c\rho^2}{\left(1 + \sqrt{1 - \epsilon c^2 \rho^2}\right)} + D_{\text{con}}(\rho/\rho_{\text{max}}) , \tag{B.58}$$

where $D_{\text{con}}(u)$ is the departure from a conic. In the case of strong aspheres it is defined by

$$D_{\text{con}}(u) := u^4 \sum_{m=0}^{M} a_m Q_m^{\text{con}}(u^2) . \tag{B.59}$$

Now one can show that it is optimal for $Q_m^{\text{con}}(x)$ to be chosen to be the Forbes Polynomials. They are defined as

$$Q_m^{\text{con}}(x) = P_m^{(0,4)}(2x - 1) \tag{B.60}$$

where $P_m^{(0,4)}$ are a particular case of the Jacobi Polynomials. The first six members are written as:

$$Q_0^{\text{con}}(x) = 1$$
$$Q_1^{\text{con}}(x) = -(5 - 6x)$$
$$Q_2^{\text{con}}(x) = 15 - 14x(3 - 2x)$$
$$Q_3^{\text{con}}(x) = -\{35 - 12x[14 - x(21 - 10x)]\}$$
$$Q_4^{\text{con}}(x) = 70 - 3x\{168 - 5x[84 - 11x(8 - 3x)]\}$$
$$Q_5^{\text{con}}(x) = -[126 - x(1260 - 11x\{420 - x[720 - 13x(45 - 14x)]\})]$$

In Fig. B.13 the first six basis elements are shown. Equation (B.58) can be reformulated, if only aspheres, that deviate mildly from a sphere, are considered. This is advantageous since manufacturability and cost effectiveness is improved for those aspheres. The best-fitting sphere in this case is chosen to be the sphere that coincides with the aspherical surface at its axial point and around its perimeter. The sag can now be expressed as

$$z(\rho) = \frac{c_{\text{bfs}}\rho^2}{\left(1 + \sqrt{1 - c_{\text{bfs}}^2 \rho^2}\right)} + D_{\text{bfs}}(\rho/\rho_{\text{max}}) \tag{B.61}$$

where the deviation is now described by

$$D_{\text{bfs}}(u) := \frac{u^2(1 - u^2)}{\sqrt{1 - c_{\text{bfs}}^2 \rho_{\text{max}}^2 u^2}} \sum_{m=0}^{M} a_m Q_m^{\text{bfs}}(u^2). \tag{B.62}$$

$D_{\text{fbs}}(u)$ is explicitly forced to vanish at the aperture's center and edge, i.e. at $u = 0$ and $u = 1$. The first three Q_m^{bfs} are given explicitly below.

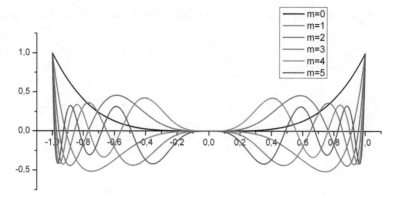

Fig. B.13 Plots of the orthogonal basis elements for m = 0, 1, 2,...5

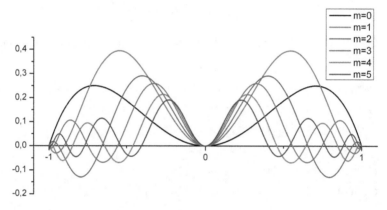

Fig. B.14 Plots for m = 0, 1, 2,...5 of the basis elements that are tailored for use when constraining the aperture from a best-fitting sphere along its normal

$$Q_0^{bfs}(x) = 1$$

$$Q_1^{bfs}(x) = \frac{1}{\sqrt{19}}(13 - 16x)$$

$$Q_2^{bfs}(x) = \sqrt{\frac{2}{95}}[29 - 4x(25 - 19x)]$$

$$Q_3^{bfs}(x) = \sqrt{\frac{2}{2545}}\{207 - 4x[315 - x(577 - 320x)]\}$$

$$Q_4^{bfs}(x) = \frac{1}{3\sqrt{131831}}(7737 - 16x\{4653 - 2x[7381 - 8x(1168 - 509x)]\})$$

$$Q_5^{bfs}(x) = \frac{1}{3\sqrt{6632213}}[66657 - 32x(28338 - x\{135325 - 8x[35884 - x(34661 - 12432x)]\})$$

A plot of the first six basis elements is shown in Fig. B.14.

B.3.10   Gaussian Optics

The electromagnetic wave, emitted by each laser, can be described by the wave equation in the scalar approximation [12]:

$$\left(\frac{\partial^2}{\partial x^2} + \frac{\partial^2}{\partial y^2} + \frac{\partial^2}{\partial z^2} - \frac{1}{c}\frac{\partial^2}{\partial t^2} \right) E(x, y, z, t) = 0. \tag{B.63}$$

Here $E(x, y, z, t)$ is the field strength and $c = c_0/n$ the speed of light in each medium, which is given by the speed of light in vacuum and the refractive index. The simplest solutions to the wave equation (B.63) are, for example, the plane wave or also the spherical waves as described in equation (B.64):

$$E(x, y, z, t) = \frac{A}{\sqrt{x^2 + y^2 + z^2}} \exp\left[-i \left(k\sqrt{x^2 + y^2 + z^2} - \omega t \right) \right]. \tag{B.64}$$

Here $A/\sqrt{x^2 + y^2 + z^2}$ is the amplitude of the field strength. These approaches, however, are unsuitable to describe the laser beam as a plane wave has an infinite expansion and thus infinite energy. A spherical wave propagates in all directions and gives also not an useful description. An improved model for a laser beam is a laterally constrained, plane wave. This results in a beam of constant intensity over the width d of an aperture cross section. According to Huygens' principle a Fraunhofer diffraction pattern arises at long distances $z > d^2/\lambda$ which diverges with increasing distance from the aperture. For a Gaussian beam profile, this divergence is at its lowest. In addition, the Gaussian intensity distribution persists at any distance. By suitable choice of the resonator of a laser one can select a Gaussian beam profile from different possible transverse field distributions or modes. This TEM_{00} fundamental mode can be regarded as a spherical wave with an imaginary wave centre. The centre of a spherical wave in equation (B.64) is arbitrary and can even be complex:

$$z \to z + iz_R = q. \tag{B.65}$$

The size z_R is real and has an important role later. By the transformation (B.65) (B.64) takes the following shape:

$$E(r, q, t) = \frac{A}{\sqrt{r^2 + q^2}} \exp\left[-i \left(k\sqrt{r^2 + q^2} - \omega t \right) \right]$$

$$\text{with } r = \sqrt{x^2 + y^2}$$

Considering the region near the z-axis ($r \ll |q|$), which is called paraxial region, this leads to:

$$E(r, q, t) \approx \frac{A}{q} \exp\left[-i\left(kq\sqrt{1 + \frac{r^2}{q^2}} - \omega t\right)\right]$$

$$\approx \frac{A}{q} \exp\left[-i\left(kq(1 + \frac{r^2}{2q^2}) - \omega t\right)\right] = \frac{B}{q} \exp\left[-i\frac{kr^2}{2q}\right] \exp[i(\omega t - kz)]$$

with $B = A \cdot \exp[kz_R]$.

The complex parameter $1/q$ can be written as follows:

$$\frac{1}{q} = \frac{1}{R(z)} - i\frac{2}{kw^2(z)}.$$

The values $w(z)$ and $R(z)$ are the spot size and radius of curvature of the wave fronts. Thus, the field distribution of a Gaussian beam is approximately (Fig. B.15)

$$E(r, z, t) \approx \frac{B}{q} \exp\left[-\frac{r^2}{w^2(z)}\right] \exp\left[-i\frac{kr^2}{2R(z)}\right] \exp[i(\omega t - kz)] \qquad (B.66)$$

This field distribution corresponds to a Gaussian function $\exp(-r^2/w^2)$ in the r direction. The spot size or beam width $w(z)$ of the TEM_{00} mode is - according to [13] - the distance $r = w(z)$ from the z-axis, in which the field $E(r, z, t)$ has decreased to $1/e$ of $E(0, z, t)$. The intensity distribution $I \sim |E|^2$ is then given by:

$$I/I_0 = \exp\left[-\frac{2r^2}{w^2(z)}\right]. \qquad (B.67)$$

Fig. B.15 Field distribution and intensity profile of a Gaussian beam

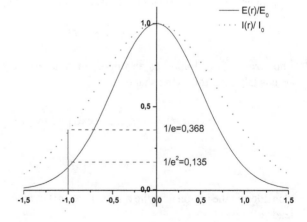

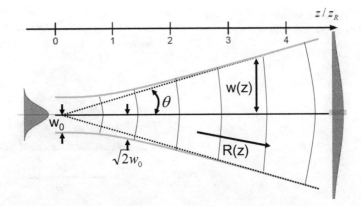

Fig. B.16 Schematic sketch of Gaussian beam with all relevant parameters

Equation (B.66) leads to the following parameters:

Beam waist:
$$w_0 = \sqrt{\frac{2z_R}{k}} = \sqrt{\frac{z_R \lambda}{\pi}}$$

Rayleigh length:
$$z_R = \frac{\pi w_0^2}{\lambda}$$

At $z = z_R$ the beam widens to $\sqrt{2}$ times the value of w_0.

Radius of curvature:
$$R(z) = z + \frac{z_R^2}{z}$$

Spot size:
$$w(z) = w_0 \sqrt{1 + (z/z_R)^2}$$

Half opening angle:
$$\theta = \frac{\lambda}{\pi w_0}$$

The radius of curvature R of the equiphase surfaces is infinite at the point of the beam waist. The behaviour of the Gaussian beam in the far field ($z \gg z_R$) is described by the angle of opening 2θ. Also in the far field the equiphase surfaces have a radius of curvature of $R = |z|$. In Fig. B.16 it can be seen that a Gaussian beam can be focused with the aid of an ideal optical system - which does not cut the beam profile - on only a minimal Beam width on the size of the beam waist. However, a Gaussian beam is cut in every optical system due to a finite size of the aperture. A Gaussian beam with a power P is given according to (B.67) and [13] by:

$$I(r) = \frac{2P}{\pi w^2} \exp\left[-\frac{2r^2}{w^2}\right].$$

In many applications, the Gaussian beam is dimmed by an aperture having a radius corresponding to the beam width (Fig. B.17). This happens, for example, when the beam falls into the rear opening of a lens.

Fig. B.17 Gaussian beam
bordered by an aperture with
a diameter $2a$

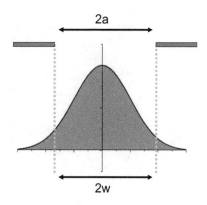

In this case, the transmitted power is given as:

$$\frac{P_{\text{Transm.}}}{P} = \frac{2}{\pi w^2} \int_0^a dr \cdot 2\pi r \cdot \exp\left[-\frac{2r^2}{w^2}\right] = 1 - \exp\left[-\frac{2a^2}{w^2}\right] \overset{a=w}{\cong} 86.47\,\%$$

(B.68)

and is described as "$1/e-$" or "86% -criterion". If the size of the aperture is $2a < 2w$, then the beam profile cover larger space than the aperture and less than 86.47% of the power will be transmitted. In such a case diffraction effects caused by the aperture (near-field Fresnel and far-field Fraunhofer diffraction effects according to [14] get more noticeable than in the case of a "filled-in" ($2a = 2w$) or "underfilled" ($2a > 2w$) aperture).

References

1. Möller, T., Trumbore, B.: Fast, minimum storage ray-triangle intersection. J. Graph. Tools **2**(1), 21–28 (1997). https://doi.org/10.1080/10867651.1997.10487468
2. Schubert, E.F.: Light-Emitting Diodes, 2nd edn. Cambridge University Press, Cambridge (2006). ISBN 9780521865388
3. Kittel, C.: Introduction to Solid State Physics. Wiley, New York (2004). http://books.google.de/books?id=kym4QgAACAAJ, ISBN 9780471415268
4. Haferkorn, H.: Optik - Physikalisch-technische Grundlagen und Anwendungen, Kapitel 1, pp. 23–163. Wiley, New York (2008). ISBN 9783527625017
5. Hunt, R.W.G., Pointer, M.: Measuring Colour. Wiley - IS&T Series in Imaging Science and Technology, 4th edn. Wiley, Chichester (2011). ISBN 1119978416
6. Kreis, T.: Holographic Interferometry: Principles and Methods. Akademie Verlag Series in Optical Metrology. Akademie Verlag (1996). http://books.google.de/books?id=qfJRAAAAMAAJ, ISBN 9783055016448
7. Science, Technology, and Applications. Version 1998. In: Bach, H., Norbert, N. (eds.) The Properties of Optical Glass. Schott Series on Glass and Glass Ceramics. Springer, Berlin (1998). https://doi.org/10.1007/978-3-642-57769-7, ISBN 978–3–642–63349–2
8. Malacara, D.: Optical Shop Testing. Wiley Series in Pure and Applied Optics. Wiley, New York (2007). http://books.google.de/books?id=qMHKB1mKFr4C, ISBN 9780470135969

9. E DIN ISO 10110-14: Optik und optische Instrumente: Erstellung von Zeichnungen für optis-
    che Elemente und Systeme, Teil 14: Toleranzen fuer Wellenfrontdeformationen (ISO/DIS
    10110-14.2:2001)
10. Withers, C.S.: Othogonal functions and Zernike polynomials - a random variable interpreta-
    tion. ANZIAM J. **50**(1), 435–444 (2009). https://doi.org/10.1017/S1446181109000169, ISSN
    1446-8735
11. Forbes, G.W.: Robust, efficient computational methods for axially symmetric optical aspheres.
    Opt. Express **18**(19), 19700–19712 (2010). https://doi.org/10.1364/OE.18.019700
12. Jackson, J.D.: Klassische Elektrodynamik. de Gruyter (1981). http://books.google.de/books?
    id=JFdwygAACAAJ, ISBN 9783110074154
13. Meschede, D.: Optik, Licht und Laser. Teubner Studienbücher. Teubner B.G. GmbH (2005).
    http://books.google.de/books?id=nGo8HKB4qCwC, ISBN 9783519132486
14. Kreis, T.: J. Opt. Soc. Am. A (6), 847–855

Index

© Springer Nature Switzerland AG 2018
S. Stuerwald, *Digital Holographic Methods*, Springer Series
in Optical Sciences 221, https://doi.org/10.1007/978-3-030-00169-8

CPSIA information can be obtained
at www.ICGtesting.com
Printed in the USA
LVHW062058031118
595631LV00032BB/28/P